SpringerBriefs in Molecular Science

For further volumes:
http://www.springer.com/series/8898

Yi-Tao Long · Chao Jing

Localized Surface Plasmon Resonance Based Nanobiosensors

 Springer

Yi-Tao Long
Chao Jing
Key Laboratory for Advanced Materials
 and Department of Chemistry
East China University of Science
 and Technology
Shanghai
China

ISSN 2191-5407   ISSN 2191-5415   (electronic)
ISBN 978-3-642-54794-2 ISBN 978-3-642-54795-9   (eBook)
DOI 10.1007/978-3-642-54795-9
Springer Heidelberg New York Dordrecht London

Library of Congress Control Number: 2014935148

Printed on acid-free paper

Springer is part of Springer Science+Business Media (www.springer.com)

Preface

Localized surface plasmon resonance (LSPR) is a phenomenon that occurs in noble-metal nanoparticles with dimensions (3–100 nm) much smaller than the wavelength of incident light (400–900 nm). The active surface electrons of nanoparticles exhibit collective oscillation upon excitation by incoming light. Because of the small size of the nanoparticles, the electrons are confined within the nanoparticle surface area and are unable to propagate. When the frequency of the incoming photons overlaps with the frequency of the electrons as they oscillate against the restoring forces of the positive nuclei, plasmon resonance is generated. Because of their LSPR properties, plasmonic nanoparticles exhibit characteristic scattering and absorption bands that make them excellent nanoprobes for a variety of applications.

In this book, we focus primarily on the plasmonic applications in biosensing and attempt to provide an overview of the different operating principles of plasmonic sensors, particularly single-nanoparticle-based detections. In Part I, we introduce the basic theory of surface plasmon resonance according to Maxwell's equations and Mie theory to explain the photon–electron interaction mechanism and resonance conditions. Part II highlights the applications of nanoplasmonics in biosensors. As the LSPR band is dependent on the particle size, shape, composition, and surrounding medium, we therefore present a series of creative biosensor designs that are based on the modulation of these nanoparticle parameters for label-free detection. Chapter 5 reviews the interparticle coupling effect, in which closely spaced nanoparticles experience LSPR coupling that causes distinct intensity and wavelength changes. Through this mechanism, distance-controlled plasmon resonance enables ultra-sensitive detection even at the single-molecule level. In Chap. 6, we summarize the discovery of plasmon resonance energy transfer and its utilization in the detection of biomolecules and heavy-metal ions. In Chap. 7, we discuss the electron transfer on plasmonics surface as well as its influence on charge separation of semiconductors. The use of nanoplasmonics in living cell imaging and photothermal therapy is emphasized in Chap. 8. Finally, in Chap. 9, we summarize the importance of LSPR and nanosensors and propose the development of future plasmonic applications.

We hope this booklet will help researchers with expertise or interest in plasmonics to gain a general understanding of its fundamental theory and applications.

Yi-Tao Long

Chao Jing

Contents

Part I
Fundamentals of Localized Surface Plasmon Resonance

Chapter 1
Brief Introduction to Localized Surface Plasmon Resonance and Correlative Devices

Abstract Novel metal nanoparticles with localized surface plasmon resonance (LSPR) have excellent optical and physical properties including strong absorption and scattering spectroscopy, photostability, and active catalytic ability. These properties enable them to be applied in variable sensitive sensors, functional nanoprobes and act as efficient catalysts. Particularly, the measurements at single nanoparticle level promote the developments of plasmonics, even to single molecule level detection. In this chapter, we briefly introduce the fundamentals and applications of the LSPR property of metal nanoparticles, and the useful devices for the investigation of single plasmonics.

Keywords Localized surface plasmon resonance • Medium sensitivity • Morphology of nanoparticles • Charge separation • Coupling of plasmonics • Electrochemistry • Cell imaging • Dark-field microscopy • Lasers

Localized surface plasmon resonance (LSPR) is the interaction between noble-metal particles' surface electrons and incident light confined on the nanoparticle surface as shown in Fig. 1.1 [1]. For nanoparticles far smaller than the wavelength of the light, the surface electrons collectively oscillate with the light transmission. When the oscillation frequency of electrons and photons is matched, the plasmon resonance occurs [2]. Due to the active LSPR band, plasmonics have unique extinction spectra including absorption and Rayleigh scattering light, as well as excellent catalytic ability benefiting from the abundant free electrons [3, 4].

In addition, plasmonics have many advantages in applications including good conductivity, light sensitivity, facile modification, and easy modulation in variable LPSR band [5, 6]. Notably, the investigation of single nanoparticles makes it possible to observe at single molecule level [7]. Thus, various sensors were constructed for label-free detection with high sensitivity and selectivity [8–10]. For instance, LSPR band is highly sensitive to the surrounding medium and displays

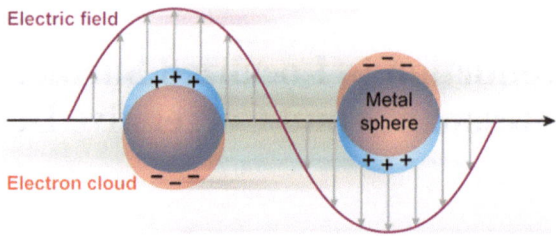

Fig. 1.1 Principle of localized surface plasmon resonance: the interaction between surface electrons and incident light. Reproduced from Ref. [1] by permission of The Royal Society of Chemistry

redshift as the medium refractive index increases. When molecules adsorb on nanoparticles, especially large proteins, the LSPR spectra changes provide a quantitative detection method [11]. Besides, as we modulate the morphology and composition of nanoparticles with different structures such as nanorods, nanoprisms, and core–shells, the resonance peak could be tuned from visible to near infrared region which are capable to be applied in wide fields [12–15]. In addition, the distance coupling of plasmonics enables detection at single molecule level with highly enhanced sensitivity. Utilising the biocompatibility and nontoxicity, gold nanoparticles play important roles in cell imaging, drug delivery, biorecognition, and photothermal therapy [16–19]. Furthermore, in combination with electrochemistry, the electron transfer and mass exchange could be monitored via the LSPR shift which is dependent on the surface electron density [20, 21]. Moreover, the development of plasmonic-semiconductor complex provides a promising new photovoltaic cell system based on the plasmon-induced charge separation [22]. Therefore, taking the advantages of plasmonics would improve the development of nanosensors in a variety of fields including life science, resources, and environment detection.

Monitoring the LSPR band, including the absorption and scattering of light, especially at single nanoparticle level, needs advanced optical techniques such as UV-Vis spectroscopy, dark-field microscopy (DFM), and differential interference contrast (DIC) microscopy [23–25]. Particularly, dark-field microscopy is a type of side illumination technique that provides a simple way to detect single nanoparticles with enhanced contrast showing a dark, even black background. For researchers who concentrate on the Rayleigh scattering light, dark-field microscopy avoids the disturbance from the illuminating light [26]. The structure of a typical dark-field condenser is shown in Fig. 1.2. Light is collected by the collector lens, and an opaque disc ("patch stop") is placed at the centre of the light route, leaving just a ring of the incident light beam. Thus, after the light passing through the condenser, two opposite "dark cones" are created. By adjusting the position of the condenser to locate the specimen on the focal point of the dark-field light route, only the light scattered by the sample can enter the objective lens, whereas

Fig. 1.2 Schematic of the experimental arrangement for dark-field microscopy studies of metal nanoparticles

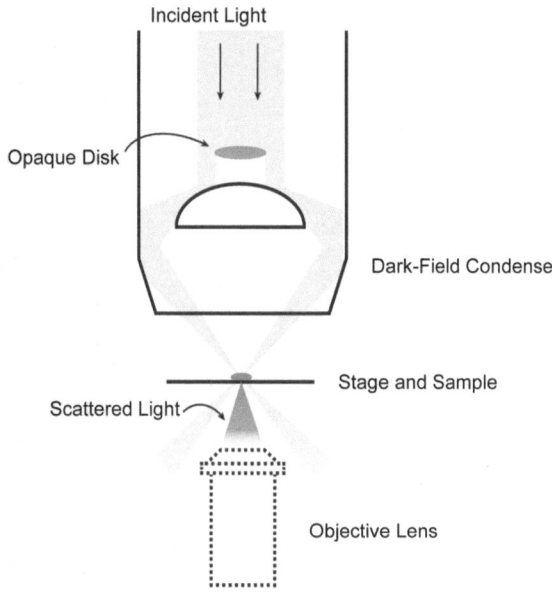

the directly transmitted light is rejected. Because of the excellent performance of DFM, it has been widely used in biological and materials sciences. Coupled with a spectrograph and a CCD camera, the Rayleigh scattering spectra from single particles can be readily measured.

The light source is another important component of the instrument. To obtain a scattering spectrum, a halogen white light source is most commonly used as the incident light in DFM. Notably, Lasers also serve as important and useful light sources, providing incident light with high-energy and narrow spectrum.

Using lasers with varying wavelengths can produce different scattering results, which is important in LSPR imaging. For instance, imaging small noble-metal plasmonic nanoparticles in a living cell is complicated due to the strong interference signals produced by organelles in the cell. He and co-workers demonstrated a new dual-wavelength difference (DWD) imaging method using a conventional dark-field microscope and lasers with different wavelengths to track gold nanoparticles in living cells [27]. Their work is based on a unique optical property of plasmonic nanoparticles: their ability to scatter light most strongly at the resonance frequency. Thus, the strongest particle signals are obtained when the wavelength of the laser matches the resonance frequency of the particles; the signal intensity will decrease substantially if the excitation light undergoes a slight shift in wavelength. Importantly, the signals from the organelles in the cell are insensitive to wavelength variations. On the basis of this phenomenon, they employed two laser beams—one as a probe beam with a wavelength of 532 nm and the other as

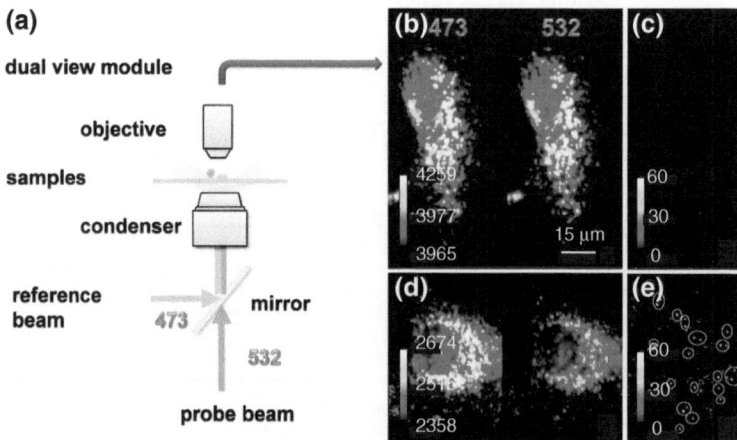

Fig. 1.3 **a** Scheme of the dual-wavelength difference (*DWD*) imaging method. Panels (**b**) and (**d**) are the conventional dark-field images of blank and GNPs-loaded HeLa cells from the 473 and 532 nm channels, respectively. Their corresponding DWD images are shown in panels (**c**) and (**e**). The GNPs on the cell membrane are highlighted with *circles*. The *scale bar* for these images is shown in panel **b**. Reprinted with the permission from Ref. [27]. Copyright 2011 American Chemical Society

a reference beam with a wavelength of 473 nm—to obtain two different images of the cell containing the gold nanoparticles (GNPs) with obvious intensity differences. Subsequently, the cellular background could be readily subtracted. With this technique, real time tracking of GNPs can be easily achieved (Fig. 1.3).

Incident light can be used not only as the source of the scattered light energy but also for other purposes, such as heating. To avoid the strong background in cells or tissues, Lounis and co-workers developed a photothermal interference contrast (PIC) technique to optically detect absorbing nanoparticles [28]. This technique benefits from the photothermal effect caused by the strong absorption of small metal particles at their plasmonic resonance. When a nanoparticle is illuminated by a laser beam that overlaps with the plasmon resonance frequency of the particle, a photothermal effect involving a change in temperature around the particle will lead to changes in the local index of refraction, which can then be detected using a second laser beam. Particles in thick samples, such as cells or tissues, can be imaged using this method. An inverted microscope with a specific design has been used to achieve this imaging (Fig. 1.4).

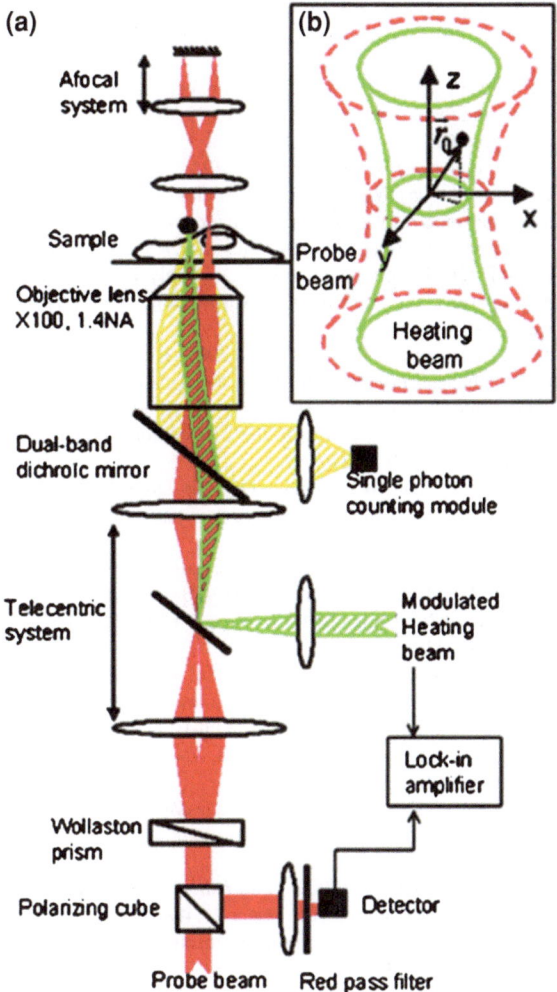

Fig. 1.4 a Schematic diagram of the optical set-up. The heating beam (in *green*) heats the metallic nanoparticle. The modulated red beam is split by a Wollaston prism into two probe beams that are retroreflected by an afocal system to be recombined by the Wollaston prism. After demodulation of one output of this interferometer, the modulated phase shift induced by the heating present on one arm of the interferometer can be detected when a nanoparticle is present. Fluorescent labels excited by the heating beam can also be detected on the same set-up by using the single-photon-counting avalanche photodiode. **b** Coordinate system with an origin at the focal point of the heating beam that coincides with one of the two red beams, the probe beam. Reprinted from Ref. [28] with permission. Copyright 2007 National Academy of Sciences USA

Furthermore, it is important to develop new techniques to investigate the absorption and scattering light of single plasmonics in different environments with various sizes to extend their applications and enhance signal intensity.

References

1. Li Y, Jing C, Zhang L, Long Y-T (2012) Resonance scattering particles as biological nanosensors in vitro and in vivo. Chem Soc Rev 41:632–642
2. Stewart ME, Anderton CR, Thompson LB, Maria J, Gray SK, Rogers JA et al (2008) Nanostructured plasmonic sensors. Chem Rev 108:494–521
3. Eustis S, El-Sayed MA (2006) Why gold nanoparticles are more precious than pretty gold: noble metal surface plasmon resonance and its enhancement of the radiative and nonradiative properties of nanocrystals of different shapes. Chem Soc Rev 35:209–217
4. Daniel M-C, Astruc D (2004) Gold nanoparticles: assembly, supramolecular chemistry, quantum-size-related properties, and applications toward biology, catalysis, and nanotechnology. Chem Rev 104:293–346
5. Jain PK, Huang X, El-Sayed IH, El-Sayed MA (2007) Review of some interesting surface plasmon resonance-enhanced properties of noble metal nanoparticles and their applications to biosystems. Plasmonics 2:107–118
6. Haruta M (2004) Gold as a novel catalyst in the 21st century: preparation, working mechanism and applications. Gold Bull 37:27–36
7. Bingham JM, Willets KA, Shah NC, Andrews DQ, Van Duyne RP (2009) Localized surface plasmon resonance imaging: simultaneous single nanoparticle spectroscopy and diffusional dynamics. J Phys Chem C 113:16839–16842
8. Haes AJ, Van Duyne RP (2004) A unified view of propagating and localized surface plasmon resonance biosensors. Anal Bioanal Chem 379:920–930
9. Sepúlveda B, Angelomé PC, Lechuga LM, Liz-Marzán LM (2009) LSPR-based nanobiosensors. Nano Today 4:244–251
10. Xia F, Zuo X, Yang R, Xiao Y, Kang D, Vallée-Bélisle A et al (2010) Colorimetric detection of DNA, small molecules, proteins, and ions using unmodified gold nanoparticles and conjugated polyelectrolytes. Proc Natl Acad Sci 107:10837–10841
11. Anker JN, Hall WP, Lyandres O, Shah NC, Zhao J, Van Duyne RP (2008) Biosensing with plasmonic nanosensors. Nat Mater 7:442–453
12. Chang W-S, Willingham B, Slaughter LS, Dominguez-Medina S, Swanglap P, Link S (2012) Radiative and nonradiative properties of single plasmonic nanoparticles and their assemblies. Acc Chem Res 45:1936–1945
13. Huang X, Neretina S, El-Sayed MA (2009) Gold nanorods: from synthesis and properties to biological and biomedical applications. Adv Mater 21:4880–4910
14. Szunerits S, Boukherroub R (2012) Sensing using localized surface plasmon resonance sensors. Chem Commun 48:8999–9010
15. Maye MM, Luo J, Han L, Kariuki NN, Zhong C-J (2003) Synthesis, processing, assembly and activation of core–shell structured gold nanoparticle catalysts. Gold Bull 36:75–82
16. Alkilany AM, Lohse SE, Murphy CJ (2012) The gold standard: gold nanoparticle libraries to understand the nano–bio interface. Acc Chem Res 46:650–661
17. Murphy CJ, Gole AM, Stone JW, Sisco PN, Alkilany AM, Goldsmith EC et al (2008) Gold nanoparticles in biology: beyond toxicity to cellular imaging. Acc Chem Res 41:1721–1730
18. Dreaden EC, Mackey MA, Huang X, Kang B, El-Sayed MA (2011) Beating cancer in multiple ways using nanogold. Chem Soc Rev 40:3391–3404
19. Lakowicz JR (2006) Plasmonics in biology and plasmon-controlled fluorescence. Plasmonics 1:5–33
20. Ling J, Huang CZ (2010) Energy transfer with gold nanoparticles for analytical applications in the fields of biochemical and pharmaceutical sciences. Anal Methods 2:1439–1447
21. Novo C, Funston AM, Gooding AK, Mulvaney P (2009) Electrochemical charging of single gold nanorods. J Am Chem Soc 131:14664–14666
22. Tian Y, Tatsuma T (2005) Mechanisms and applications of plasmon-induced charge separation at TiO_2 films loaded with gold nanoparticles. J Am Chem Soc 127:7632–7637

23. Augspurger AE, Stender AS, Han R, Fang N (2014) Detecting plasmon resonance energy transfer with differential interference contrast microscopy. Anal Chem 86:1196–1201

24. Rosi NL, Mirkin CA (2005) Nanostructures in biodiagnostics. Chem Rev 105:1547–1562

25. Otte MA, Sepúlveda B, Ni W, Juste JP, Liz-Marzán LM, Lechuga LM (2009) Identification of the optimal spectral region for plasmonic and nanoplasmonic sensing. ACS Nano 4:349–357

26. Hu M, Novo C, Funston A, Wang H, Staleva H, Zou S et al (2008) Dark-field microscopy studies of single metal nanoparticles: understanding the factors that influence the linewidth of the localized surface plasmon resonance. J Mater Chem 18:1949–1960

27. Xiao L, Wei L, Cheng X, He Y, Yeung ES (2011) Noise-free dual-wavelength difference imaging of plasmonic resonant nanoparticles in living cells. Anal Chem 83:7340–7347

28. Cognet L, Tardin C, Boyer D, Choquet D, Tamarat P, Lounis B (2003) Single metallic nano-particle imaging for protein detection in cells. Proc Natl Acad Sci 100:11350–11355

Chapter 2
Electromagnetics of Metals and Theory Fundamentals

Abstract In this chapter, we focus on describing the classical electromagnetic theory to expound the basic fundamentals of small discrete particles. With this theory, we can finally derive the mechanism and property of plasmonic particles like gold, silver, copper, etc. Here, we first lead in the Maxwell's equations. They will make it easier to understand the interaction of electromagnetism. Then, we introduce some fundamental parameters, such as the refractive index and the relative parameters, to make the analysis clearly. These quantities will be used in the simplest plane-wave solution and surface plasmon polariton (SPPS) equation which are the intermediate processes that can lead us to the final answer. In order to explain the mechanism clearly, we do not want to twist in the objective factor, so the light we use here is a monochromatic parallel beam, and dipole excitation is considered as the most important factor. And the light containing the general properties of optics can ignore plenty of special cases.

Keywords Maxwell's equation • Plane wave • Refractive index • Relative parameters • Dipole excitation • Adsorption and scattering

2.1 Maxwell's Equations

In space, the physical electromagnetic field is generated by moving electrically charged objects. When the objects move, the electric and magnetic fields change at the same time. So, it can be easily realised that we can use these two vectors, E and B, to represent electromagnetic field, called the electric vector and the magnetic induction, respectively.

As we know, when the charged objects move quickly, the charges can form the electric current. It is naturally to deduce the electric current density j, the electric displacement D, and the magnetic vector H. Unless otherwise stated, the particles used here are noble metals.

Y.-T. Long and C. Jing, *Localized Surface Plasmon Resonance Based Nanobiosensors*,
SpringerBriefs in Molecular Science, DOI: 10.1007/978-3-642-54795-9_2,
© The Author(s) 2014

These are the original quantities that can finally derive Maxwell's equations with the space and time derivatives [1].

$$\text{curl } H - \frac{1}{c}D' = \frac{4\pi}{c}j \tag{2.1}$$

$$\text{curl } E + \frac{1}{c}B' = 0 \tag{2.2}$$

$$\text{div } D = 4\pi\rho \tag{2.3}$$

$$\text{div } B = 0 \tag{2.4}$$

It is worthwhile to note that the dot here shows differentiation with respect to time, and the physical property of medium at any point is continuous. The constant c in (2.1) and (2.2) is the velocity of light in vacuum and is approximately equal to 3×10^8 m/s.

From the derivation of (2.1), we can get [2]

$$\text{div } j = -\frac{1}{4\pi}\text{div } D' \tag{2.5}$$

Since div curl $\equiv 0$, and use (2.3), we can get

$$\frac{\partial\rho}{\partial t} + \text{div } j = 0 \tag{2.6}$$

When the charge is conserved at any point of the medium, it can be integrated under the condition that Gauss' theorem is fitted:

$$\frac{d}{dt}\int\rho\,dV + \int j\cdot n\,dS = 0 \tag{2.7}$$

The first integral is taking place throughout the volume, and the second one is bounded in the surface region, n denotes the unit outward normal. This equation implies that the total charge

$$e = \int\rho\,dV \tag{2.8}$$

The total charge e in Eq. (2.8) contained within the domain can only increase on account of the flow of electric current, and then, we can get

$$J = \int j\cdot n\,dS \tag{2.9}$$

The situation introduced here is dependent on the condition that the field quantities are independent to time. If time is discontinuous, what will happen? It can be imagined that the charged objects may move and appear anywhere. If there is no current ($j = 0$), the field is called a static field; if current exists ($j \neq 0$), the field is called a stationary field.

We use material equations (or constitutive relations) to define the relations of substances in the physical field. The relations that describe the behaviour of substances under the influence of the field are called material equations (or constitutive relations). As we know, it is complicated and difficult to analyse. Here, we just discuss the relations when the field is time harmonic and the physical properties of it are independent of location, direction, and angle. So, it is easy to form the relations [3, 4]:

$$j = \sigma E \tag{2.10}$$

$$D = \varepsilon E \tag{2.11}$$

$$B = \mu H \tag{2.12}$$

Here, σ is called the specific conductivity, ε is known as the dielectric constant (or permittivity), and μ is called the magnetic permeability.

2.2 The Wave Equation

In order to interpret the relationship between the field vectors and the differential equations better, we focus our attention on the part where $j = 0$ and $\rho = 0$.

We put B from (2.2) into (2.12). After the deformation of the equation, it gives [2]

$$\text{curl} \left(\frac{1}{\mu} \text{curl } E \right) + \frac{1}{c} \text{curl } H' = 0 \tag{2.13}$$

Take the differential quotient to time of (2.1); substitute E for D using (2.11). Then, it gives

$$\text{curl} \left(\frac{1}{\mu} \text{curl } E \right) + \frac{\varepsilon}{c^2} E'' = 0 \tag{2.14}$$

Be aware of the identities $\text{curl } uv = u \cdot \text{curl } v + (\text{grad } u) \times v$ and $\text{curl curl} = \text{grad div} - \nabla^2$ (2.14) becomes [5]

$$\nabla^2 E - \frac{\varepsilon \mu}{c^2} E'' + (\text{gard ln } \mu) \times \text{curl } E - \text{gard div } E = 0 \tag{2.15}$$

With the help of the identity, $\text{div } uv = u \cdot \text{div } v + v \cdot \text{grad } u$ (2.3) becomes

$$\varepsilon \cdot \text{div } E + E \cdot \text{gard } \varepsilon = 0 \tag{2.16}$$

After collating (2.15), it gives

$$\nabla^2 E - \frac{\varepsilon \mu}{c^2} E'' + (\text{gard ln } \mu) \times \text{curl } E + \text{gard}(E \cdot \text{gard ln } \varepsilon) = 0 \tag{2.17}$$

We can also get the similar answer if we substitute H for B using (2.12) and (2.13) gives

$$\nabla^2 H - \frac{\varepsilon \mu}{c^2} H'' + (\text{gard ln } \varepsilon) \times \text{curl } H + \text{gard}(H \cdot \text{gard ln } \mu) = 0 \tag{2.18}$$

It is worth mentioning that if the medium is homogeneous, i.e. gard $\log \varepsilon =$ gard $\ln \mu = 0$ (2.17) and (2.18) become

$$\nabla^2 E - \frac{\varepsilon\mu}{c^2}E'' = 0 \tag{2.19}$$

$$\nabla^2 H - \frac{\varepsilon\mu}{c^2}H'' = 0 \tag{2.20}$$

The equations prove that the electromagnetic waves surely exist, and they are standard equations of wave motion. And we can get the velocity in the medium

$$v = \frac{c}{\sqrt{\varepsilon\mu}} \tag{2.21}$$

Compare (2.21) with $n = \frac{c}{v}$, we can get [6]

$$n = \sqrt{\varepsilon\mu} \tag{2.22}$$

For the most transparent substance, μ is practically equal to unity, n is the refractive index. That is why we consider ε as a constant normally. When we take the atomic structure of matter into account, ε is not an immutable characteristic of the material. It will be discussed in the next section.

2.3 Scattering from Inhomogeneous Media

Scattering is a physical process in which light is deflected haphazardly as a result of collisions. Here, we shall confine our attention to the part that the relationship between media and the incident wave is linear. When the physical property is time independent, it is called static scattering.

At first, we will take two extreme media into consideration. One is conducting medium, the other is nonconducting medium.

For the conducting one, we assume that it is not related to V, the space-dependent part in Eq. (2.17) can be extracted from the complex electric field, deform it in the monochromatic electromagnetic field, we can get [7]

$$\nabla^2 E(r,\omega) + k^2\varepsilon(r,\omega)E(r,\omega) + \text{gard}\left[E(r,\omega)\cdot\text{gard ln }\varepsilon(r,\omega)\right] = 0 \tag{2.23}$$

As we know, the medium is isotropic and nonmagnetic. So, the time-dependent part can be defined as

$$E = E_0 e^{-i\omega t} \tag{2.24}$$

(Not to explain in the following analysis), substitute (2.24) for E in (2.23), we can get

$$k^2 = \frac{\omega^2}{c^2}\left(1 + i\frac{\sigma}{\varepsilon_0\omega}\right) \tag{2.25}$$

We define $1 + i\frac{\sigma}{\varepsilon_0\omega}$ as the complex permittivity $\varepsilon_r(\omega)$. It establishes a relationship between the specific conductivity and the dielectric constant. In order to get ε, we now introduce the free-electron gas (FEG) model. It is a special quantum mechanics model that describes the property of Fermi systems in the ideal gas. It ignores the interaction between centron and electron, electron and electron, and assumes that the electrons can move freely in the background full of protons.

In the monochromatic electromagnetic field [8], motion of electrons can be shown as

$$mv' = eE \tag{2.26}$$

Compare with (2.24), we can get

$$v = \frac{-eE}{im\;\omega} \tag{2.27}$$

Since $j = nev$, substitute j for v, it becomes

$$j = -\frac{ne^2}{im\omega}E \tag{2.28}$$

The comparison of (2.10) and (2.28) gives

$$\sigma = -\frac{ne^2}{im\omega} \tag{2.29}$$

Put σ into Eq. (2.25), we can get

$$\varepsilon_r(\omega) = 1 - \frac{ne^2}{\varepsilon_0 m\omega^2} = 1 - \frac{\omega_p^2}{\omega^2} \tag{2.30a}$$

$$\omega_p^2 = \frac{ne^2}{\varepsilon_0 m} \tag{2.30b}$$

For the nonconducting one, we can also get the similar answer from Lorenz model in the dielectric. The electron theory assumes that:

First, the charges in the atoms or molecules are spread in the surface under the quasi-elastic force.

Second, dielectric will be polarised under the effect of incident light, following forced vibration of charged particle.

Third, atomic nucleus is considered to be immobile, and the charges vibrate at the inherent vibration frequency.

Since the restoring force $f = -m\omega_0^2 r$, we can easily infer the equation of electron forced vibration

$$mr'' = -eE - fr - gr' \tag{2.31}$$

Here, eE is the electric field force, fr is the quasi-elastic force, gr' is the damping force. And we lead in two quantities, eigentone $\omega_0 = \sqrt{f/m}$, damping factor $\gamma = \sqrt{g/m}$, to make the equation easy to settle (2.31) becomes

$$r'' + \gamma r' + \omega_0^2 r = -\frac{eE_0 e^{-i\omega t}}{m} \tag{2.32}$$

We then can get the solution of Eq. (2.32)

$$r(t) = \frac{(-e)}{m} \frac{E_0 e^{-i\omega t}}{(\omega_0^2 - \omega^2) - i\gamma\omega} \tag{2.33}$$

From the definition of the electric dipole moment and polarisation, we can get

$$p = qr \tag{2.34}$$

$$P = Np = -Ner \tag{2.35}$$

Put (2.33) into (2.35), we can get

$$P = \frac{Ne^2}{m} \frac{E_0 e^{-i\omega t}}{(\omega_0^2 - \omega^2) - i\gamma\omega} \tag{2.36}$$

Since $p = \varepsilon_0 \chi E$, then

$$\chi = \omega_p^2 \frac{1}{(\omega_0^2 - \omega^2) - i\gamma\omega} \tag{2.37}$$

And the complex permittivity becomes [9]

$$\varepsilon_r(\omega) = 1 + \omega_p^2 \frac{1}{(\omega_0^2 - \omega^2) - i\gamma\omega} \tag{2.38}$$

To sum up, we can make the Eq. (2.38) simply when $\omega_p \gg \omega_0$, and this is the famous Drude model

$$\varepsilon_r(\omega) = 1 - \frac{\omega_p^2}{\omega^2 + i\gamma\omega} \tag{2.39}$$

2.4   Localized Surface Plasmons

Localized surface plasmon resonance (LSPR) is different from surface polaritons, which can exist not only in the boundary of dielectric and metal, but also in the structures. LSPR frequency can be obtained by the Laplace equation.

Assuming that there is a metal particle in the vacuum, we can get its space distribution of potential through Laplace equation [10]:

$$\phi_1(r, \theta, \varphi) = \sum_{l=0}^{\infty} \sum_{m=-1}^{\infty} a_{lm} r^l Y_{lm}(\theta, \varphi) \quad 0 < r \le R \tag{2.40a}$$

Fig. 2.1 Different kinds of excitation

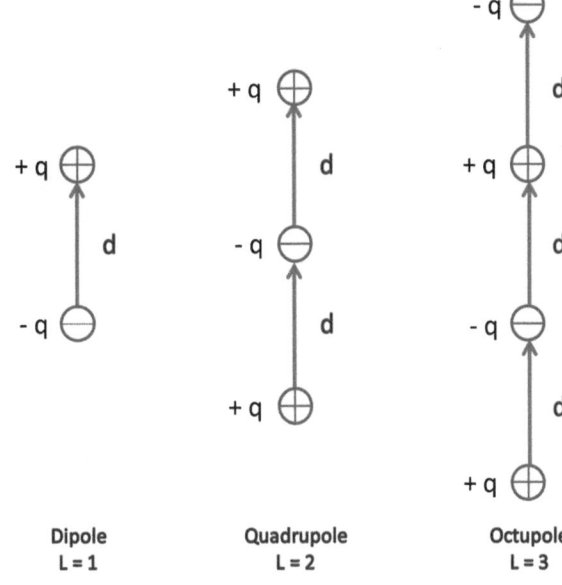

In the electrostatic approach, we can change the form of (2.40a, b) due to the azimuthal symmetry

$$\phi_2(r,\theta,\varphi) = \sum_{l=0}^{\infty} \sum_{m=-1}^{\infty} b_{lm} \frac{1}{r^{l+1}} Y_{lm}(\theta,\varphi) \quad r > R \tag{2.40b}$$

The boundary conditions: ϕ and $\varepsilon_i \frac{\partial \phi}{\partial x}$ are continuous when $r = R$. Sometimes we will consider the fact that there are other kinds of excitation like quadrupole, octupole, and so on (Fig. 2.1). We can also put them in the dipolar model. It shows

$$\omega_l = \omega_p \left(\frac{l}{2l+1} \right)^{0.5} \quad l = 1, 2, 3 \ldots \tag{2.41}$$

In this book, our conclusion is based on the simple quasi-static approximation. It suggests that the particles we use here are much smaller than the wavelength of light. Dipole excitation is the most important one [11].

In the electrostatic approach, we can change the form of (2.40a, b) due to the azimuthal symmetry

$$\phi_1(r,\theta) = \sum_{l=0}^{\infty} \left[A_l r^l + B_l r^{-(l+1)} \right] P_1(\cos\theta) \tag{2.42}$$

Here, $P_1(\cos\theta)$ are the Legendre polynomials of order 1, and θ is the angle between the position vector r at the line P and z-axis (Fig. 2.2).

In order to better understand the potentials, we excise (2.42) into ϕ_{in} and ϕ_{out}, defined as the function inside and outside the sphere, respectively, then we can get [12]

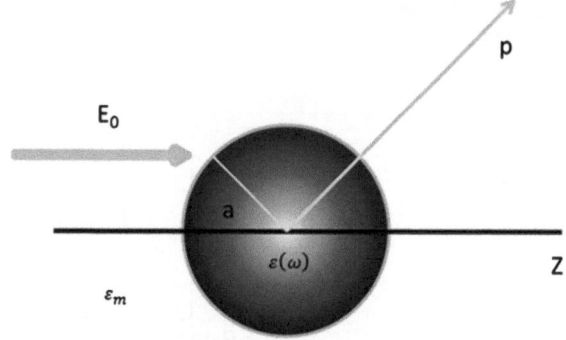

$$\phi_{\text{in}}(r,\theta) = \sum_{l=0}^{\infty} A_l r^l P_l(\cos\theta) \tag{2.43a}$$

$$\phi_{\text{out}}(r,\theta) = \sum_{l=0}^{\infty} \left[B_l r^l + C_l r^{-(l+1)} \right] P_l(\cos\theta) \tag{2.43b}$$

The boundary condition is $r \rightarrow \infty$, and then, the coefficient B_l can be determined. The requirement that $\phi_{\text{out}} \rightarrow -E_0 r \cos\theta$ as $r \rightarrow \infty$ demands that $B_l = -E_0$ and $B_l = 0$ for $l \neq 1$. The remained A_l, and C_l are defined by $r = a$. The equality of the normal components of the displacement field is

$$-\varepsilon_0 \varepsilon \frac{\partial \phi_{\text{in}}}{\partial r}\bigg|_{r=a} = -\varepsilon_0 \varepsilon_{\text{m}} \frac{\partial \phi_{\text{out}}}{\partial r}\bigg|_{r=a} \tag{2.44}$$

And the equality of the tangential components of the electric field is

$$-\frac{1}{a} \frac{\partial \phi_{\text{in}}}{\partial r}\bigg|_{r=a} = -\frac{1}{a} \frac{\partial \phi_{\text{out}}}{\partial r}\bigg|_{r=a} \tag{2.45}$$

The boundary mentioned before leads to $A_l = C_l = 0$ for $l \neq 1$, and then, we can calculate the evaluation of (2.43a, b)

$$\phi_{\text{in}} = -\frac{3\varepsilon_{\text{m}}}{\varepsilon + 2\varepsilon_{\text{m}}} E_0 r \cos\theta \tag{2.46a}$$

$$\phi_{\text{out}} = -E_0 r \cos\theta + \frac{\varepsilon - \varepsilon_{\text{m}}}{\varepsilon + 2\varepsilon_{\text{m}}} E_0 a^3 \frac{\cos\theta}{r^2} \tag{2.46b}$$

We can put the dipole moment p in (2.46b), it becomes

$$\phi_{\text{out}} = -E_0 r \cos\theta + \frac{pr}{4\pi \varepsilon_0 \varepsilon_{\text{m}} r^3} \tag{2.47}$$

$$p = 4\pi \varepsilon_0 \varepsilon_m a^3 \frac{\varepsilon - \varepsilon_m}{\varepsilon + 2\varepsilon_m} \tag{2.48}$$

According to (2.36), we can get

$$\chi = 4\pi a^3 \frac{\varepsilon - \varepsilon_m}{\varepsilon + 2\varepsilon_m} \tag{2.49}$$

We can get the same functional form as the Clausius–Mossotti relation. The extinction cross section [13], C_{ext} is easily calculated as the sum of the absorption and scattering cross section via [2]

$$C_{ext} = C_{abs} + C_{sca} \tag{2.50}$$

where C_{abs} and C_{sca} are obtained from the polarisability χ in (2.49), it gives

$$C_{sca} = \frac{k^4}{6\pi}|\chi|^2 = \frac{8\pi}{3}k^4 a^6 \left|\frac{\varepsilon - \varepsilon_m}{\varepsilon + 2\varepsilon_m}\right|^2 \tag{2.51a}$$

$$C_{abs} = k Im[\chi] = 4\pi k a^3 Im\left[\frac{\varepsilon - \varepsilon_m}{\varepsilon + 2\varepsilon_m}\right] \tag{2.51b}$$

When the size of particles is much smaller than the incident wavelength ($a \ll \lambda$), the efficiency of absorption scales with a^3 and the scattering efficiency scales with a^6. Also, the cross sections in (2.51a, b) can be applied in the expression of dielectric scatters. As $C_{sca} \propto a^6$, and that is why we cannot see the small particles in the camera like CCD, the background of scattering is too large to see small objects. We can also see that both absorption and scattering are enhanced when dipole excitation exists in the form of (2.51a, b). When we add a quasi-static limit in the dielectric function $\varepsilon = \varepsilon_1 + i\varepsilon_2$, Eqs. (2.51a, b) becomes [14–16]

$$C_{ext} = 9\frac{\omega}{c}\varepsilon_m^{1.5} V \frac{\varepsilon_2}{(\varepsilon_1 + 2\varepsilon_m)^2 + \varepsilon_2^2} \tag{2.52}$$

Here, we know how to deal with a spherical nanoparticle. However, most of the nanoparticles are not sphere, like ellipsoid or spheroid. If we suppose that the nanoparticle is an ellipsoid with axes $a \le b \le c$, then we can get $\frac{x^2}{a^2} + \frac{y^2}{b^2} + \frac{z^2}{c^2} = 1$. So, the expression of χ becomes
$L_{x,y,z}$ is called the depolarisation, which gives

$$\chi_{x,y,z} = \frac{4\pi abc(\varepsilon_{Au} - \varepsilon_m)}{3\varepsilon_m + 3L_{x,y,z}(\varepsilon_{Au} - \varepsilon_m)} \tag{2.53}$$

$$L_x = \frac{1 - e^2}{e^2}\left(-1 + \frac{1}{2e}\ln\left(\frac{1 + e}{1 - e}\right)\right) \tag{2.54}$$

$$L_{y,z} = \frac{1 - L_x}{2} \tag{2.55}$$

It should be emphasised that the above equations are approximate calculation, which is based on the electromagnetic interactions on the dipolar model situated at the centres of each particle, other kinds of excitations like quadrupole, octupole effects are ignored. The incident wave and media should also be considered as linear. The subtle relationship between surface charges is still unclear, thus, the exploring of the theory for surface plasmon resoance is crucial in understanding the interaction between photons and electrons.

References

1. Maxwell JC (1865) A dynamical theory of the electromagnetic field. Phiol Trans R Soc London 155:459–512
2. Born M, Wolf E (1999) Principles of optics. Cambridge University Press
3. Mishchenko MI, Travis LD, Lacis AA (2002) Scattering, absorption, and emission of light by small particles. Cambridge University Press
4. Mackay TG, Lakhtakia A (2010) Eletromagnetic anisotropy and bianisotropy: a field guide. World Scientific Press
5. Costabel M, Dauge M (1999) Maxwell and Lame eigenbalues on polyhedra. Math Meth App Sci 22:3–20
6. Hecht E (2002) Optics. Addison-Wesley Press
7. Kelly KL, Coronado E, Zhao LL, Schatz GC (2003) The optical properties of metal nanoparticles: the influence of size, shape, and dielectric environment. J Phys Chem B 107:668–677
8. Blaber MG, Henry AI, Bingham JM, Schatz GC, Duyne RPV (2012) LSPR imaging of silver triangular nanoprisms: correlating scattering with structure using electrodynamics for plasmon lifetime analysis. J Phys Chem C 116:393–403
9. Hu M, Novo C, Funston A, Wang H, Staleva H, Zou SL, Mulvaney P, Xiae Y, Hartland GV (2008) Dark-field microscopy studies of single metal nanoparticles: understanding the factors that influence the linewidth of the localized surface plasmon resonance. J Mater Chem 18:1949–1960
10. Hohenester U, Krenn J (2005) Surface plasmon resonances of single and coupled metallic nanoparticles: a boundary integral method approach. Phys Rev B 72:195429–195438
11. Link S, El-Sayed MA (2000) Shape and size dependence of radiative, non-radiative and photothermal properties of gold nanocrystals. Int Rev Phys Chem 19(3):409–453
12. Maier SA (2007) Plasmonics: fundamentals and applications. Springer Press
13. Jain PK, Lee KS, El-Sayed IH, El-Sayed MA (2006) Calculated absorption and scattering properties of gold nanoparticles of different size, shape, and composition: applications in biological imaging and biomedicine. J Phys Chem B 110:7238–7248
14. Ringe E, Zhang J, Langille MR, Mirkin CA, Marks LD, Duyne RPV (2012) Correlating the structure and localized surface plasmon resonance of single silver right bipyramids. Nanotechnology 23:444005–444011
15. Härtling T, Alaverdyan Y, Wenzel MT, Kullock R, Käll MM, Eng L (2008) Photochemical tuning of plasmon resonances in single gold nanoparticles. J Phys Chem C 112:4920–4924
16. Link S, El-Sayed MA (1999) Size and temperature dependence of the plasmon absorption of colloidal gold nanoparticles. J Phys Chem B. 103:4212–4217

Part II
Sensing Applications of Plasmonic Nanoparticles

Part II

Chapter 3
Refractive Index-based Plasmonic Biosensors

Abstract Plasmonic nanoparticles are very sensitive to changes in the medium surrounding their surfaces. Pronounced redshifts in the LSPR spectral position occur when the refractive index of the surrounding medium increases. Thus, the modification or adsorption of molecules on the nanoparticle surface can be detected in a label-free manner using LSPR shifts. In this chapter, we summarise the previous reports of medium- and substrate-influenced plasmon resonance and describe innovative biosensors that show high sensitivity and selectivity.

Keywords Refractive index • Effect of surrounding medium • Effect of substrate • Effect of solvent • Effect of ligands • Biosensors

3.1 Effect of Surrounding Medium on Single Nanoparticles

The effects of medium refractive index on plasmonics LSPR bands have been investigated for many years [1, 2]. In particular, monitoring the spectral changes of single nanoparticles provides more detailed and accurate information compared with measurements of bulk solutions [3, 4]. In 2003, Schultz investigated the influence of the surrounding refractive index on single silver nanoparticles in different oils [5]. As shown in Fig. 3.1, after the addition of oil, the colour of the silver nanoparticles changed from blue to green, indicating redshifts in the scattering spectra. The colour of the nanoparticles changed back to blue when the oils were removed. When the refractive index of the oil was increased, the scattering spectra gradually redshifted, and a linear relationship was observed between the peak wavelength and the refractive index of the medium.

In the same year, Van Duyne used silver nanoparticles as probes to detect the formation of a monolayer on the surface of the nanoparticles. The scattering spectra of single silver nanoparticles in different solvents, including nitrogen, methanol, 1-propanol, chloroform, and benzene, were determined, as shown in Fig. 3.2a [6].

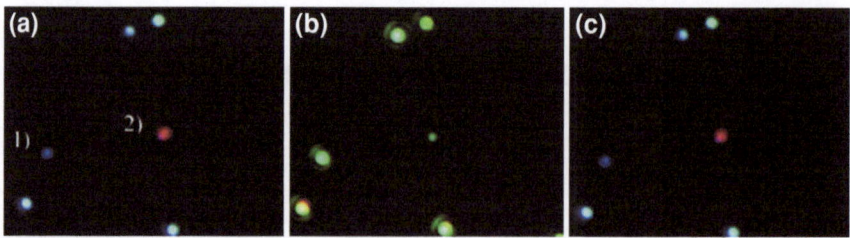

Fig. 3.1 Typical field of silver nanoparticles immobilised on SiO_2 wafer and imaged under dark-field illumination with 100× objective. *Colour images* taken with Nikon Coolpix 995. **a** Before oil. **b** With 1.44 index oil spread over particles. **c** After removal of 1.44 index oil with the described protocol. Reprinted with permission from Ref. [5]. Copyright (2003) American Chemical Society

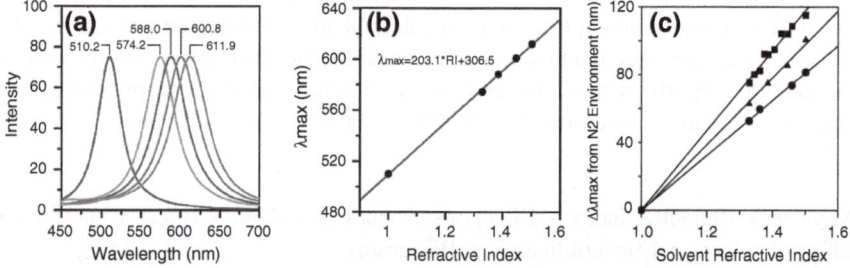

Fig. 3.2 **a** Single Ag nanoparticle resonant Rayleigh scattering spectrum in various solvent environments (*left* to *right*): nitrogen, methanol, 1-propanol, chloroform, and benzene. **b** Plot depicting the linear relationship between the solvent refractive index and the LSPR λ_{max}. **c** Comparison of refractive index sensitivity for Ag nanoparticles with different geometries. The spherical nanoparticle (*filled circle*) has a sensitivity of 161 nm RIU^{-1}, the triangular nanoparticle (*filled triangle*) has a sensitivity of 197 nm RIU^{-1}, and the rod-like nanoparticle (*filled square*) has a sensitivity of 235 nm RIU^{-1}. Adapted with permission from Ref. [6]. Copyright (2003) American Chemical Society

When the refractive index of the solvent increased, the peaks in the scattering spectra clearly redshifted in a linear manner, in agreement with previous reports. Van Duyne proposed an equation to estimate the linear relationship between the surface refractive index and the LSPR peak shift, as shown in Eq. (3.1) [7]:

$$\Delta\lambda_{max} \approx m(\Delta n)(1 - \exp(-2d/l_d)) \tag{3.1}$$

where m is the sensitivity factor of the GNPs ($\Delta\lambda_{max}$ per refractive index unit (RIU) change, nm/RIU), Δn is the refractive index change of the surrounding medium, d is the effective thickness of the adsorption layer, l_d is the electromagnetic field decay length, ε_m is the dielectric constant of the surrounding environment, and ε is the dielectric constant of the GNPs.

Notably, the structural properties of the nanoparticles influence their scattering spectra [6]. In this work, the relationship between the solvent refractive index and

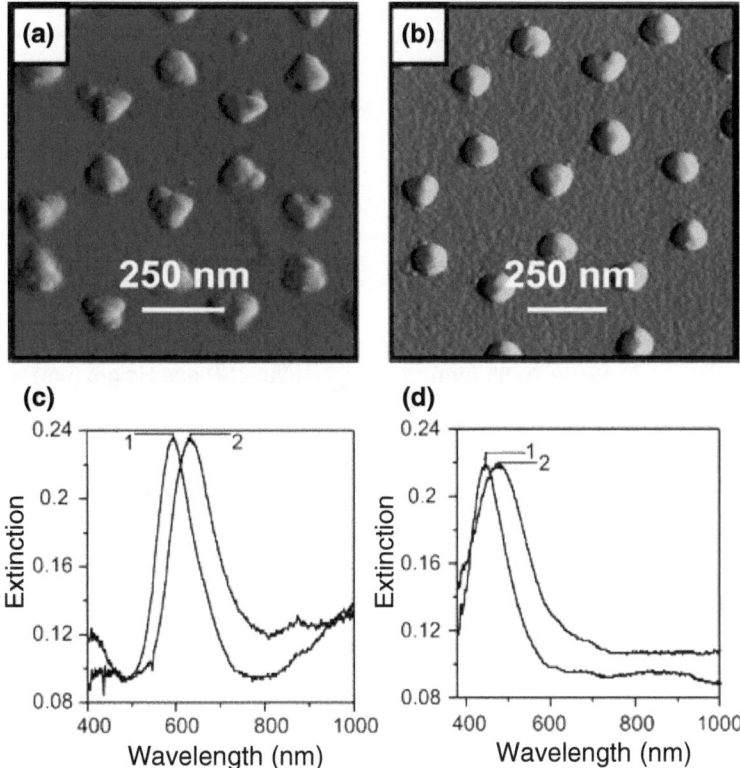

Fig. 3.3 **a–b** Representative tapping-mode AFM images of Ag nanoparticles (nanosphere diameter, $D = 400$ nm, $d_m = 50.0$ nm). Scan areas: ~1 μm². Scan rate = 2.0 Hz. **a** After solvent annealing, the resulting nanoparticles have in-plane widths $a = 114$ nm and out-of-plane widths $b = 54$ nm. **b** After thermal annealing of the sample at 650 °C for 1 h, the resulting nanoparticles have in-plane widths of ~110 nm and out-of-plane heights of ~61 nm. **c–d** LSPR spectra of nanoparticles ($a = 100$ nm, $b = 50.0$ nm) in an N₂ environment. **c** Ag nanoparticles (*1*) before chemical modification, $\lambda_{max} = 594.8$ nm, and (*2*) after modification with 1 mM hexadecanethiol, $\lambda_{max} = 634.8$ nm. **d** Annealed Ag nanoparticles (*1*) before chemical modification, $\lambda_{max} = 438.8$ nm, and (*2*) after modification with 1 mM hexadecanethiol, $\lambda_{max} = 466.4$ nm. Adapted with permission from Ref. [8]. Copyright (2004) American Chemical Society

nanoparticles with various shapes and surface modifications was explored, as depicted in Fig. 3.2c. Spherical, triangular, and rod-like nanoparticles exhibited different sensitivities to the surrounding refractive index, with sensitivities (*m*) of 161, 197, and 235 nm/RIU, respectively. Nanoparticles with higher aspect ratio are more sensitive to the surrounding medium.

In addition, the size, shape, and composition dependence of the plasmon resonance sensitivity were observed using nanoparticles produced by nanosphere lithography (NSL) [8]. The NSL-fabricated nanoparticles exhibited sensitivities similar to those of the single silver nanoparticles. Figure 3.3a and b displays AFM images of the NSL-fabricated silver nanoparticles before and after annealing

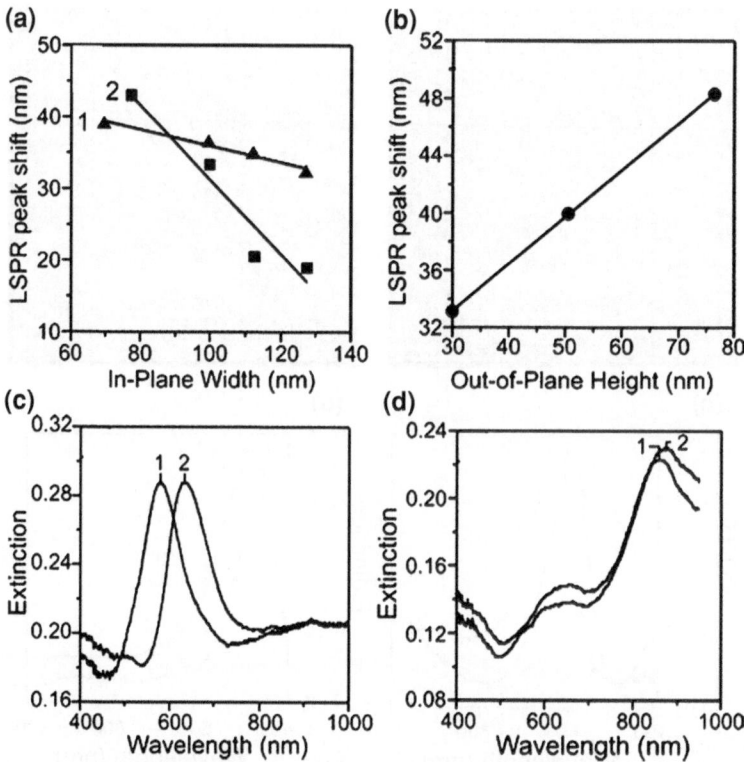

Fig. 3.4 **a–b** Hexadecanethiol LSPR shift dependence on Ag nanoparticle size. **a** (*1*) Ag nan-oparticles with a fixed out-of-plane height, $b = 50.0$ nm, and varying in-plane width. Linear regression was used to fit the data to a line described by $y = -0.1135x + 47.3$. (*2*) Ag nano-particles with a fixed out-of-plane height, $b = 30.0$ nm, and varying in-plane width. Linear regression was used to fit the data to a line described by $y = -0.52x + 83.5$. **b** Ag nanoparti-cles with a fixed in-plane width, $a = 100$ nm, and varying out-of-plane height. Linear regres-sion was used to fit the data to a line described by $y = 0.33x + 23.6$. **c–d** LSPR spectra of nanoparticles ($a = 100$ nm, $b = 75.1$ nm) in a N_2 environment. **c** Ag nanoparticles (*1*) before chemical modification, $\lambda_{max} = 563.0$ nm, and (*2*) after modification with 1 mM hexadecanethiol, $\lambda_{max} = 611.1$ nm. **d** Au nanoparticles (*1*) before chemical modification, $\lambda_{max} = 860.7$ nm, and (*2*) after modification with 1 mM hexadecanethiol, $\lambda_{max} = 874.5$ nm. Adapted with permission from Ref. [8]. Copyright (2004) American Chemical Society

at 650 °C for 1 h. After modification with 1 mM hexadecanethiol, the scattering spectra of the as-fabricated nanoparticles indicated a redshift from 594.8 to 634.8 nm, which was attributed to an adsorbate layer with a thickness of 2.61 nm. The peaks in the scattering spectra of the nanoparticles blueshifted from 594.8 to 438.8 nm after they were annealed and redshifted by approximately 27 nm after they were exposed to hexadecanethiol. The smaller redshift was due to the less oblate structure of annealed nanoparticles without sharp points. Subsequently, the LSPR spectra of nanoparticles with different widths and heights were monitored after modification with hexadecanethiol, as shown in Fig. 3.3c and d. The LSPR peak shift decreased approximately 1.1 nm per 10 nm increase in the width of the nanoparticles for a constant height of 50 nm. Similarly, for 30 nm tall silver nanoparticles, the LSPR peak shift decreased approximately 5.2 nm per 10 nm increase in the width of the nanoparticles. Nanoparticles with fixed widths and various heights are shown in Fig. 3.4. As the height of particles increased by 10 nm, the LSPR shift increased by approximately 3.3 nm. These results indicate that LSPR peak shifts decrease as the aspect ratios of the nanoparticles increase. In addition, the LSPR shifts did not display a clear relationship with the surface area and volume. In terms of nanoparticle composition, as shown in Fig. 3.4c and d, the LSPR spectra of silver and gold nanoparticles treated with hexadecanethiol suggest that silver is more sensitive than gold to changes in the surrounding refractive index.

3.2 Effects of Nonlinear Solvents and Ligands

Pal and co-workers investigated the effects of solvents and modified ligands on the plasmon resonance bands of cetylpyridinium chloride (CPC)-stabilised gold organosols in toluene [9]. When the nanoparticles were exposed to different solvents with refractive indices that ranged from 1.3 to 1.5, the majority of the spectra displayed an obvious redshift as the refractive index increased, as shown in Fig. 3.5a. However, in some cases, the relationship between λ_{max} and n was nonlinear. The linear results for cyclohexane, chloroform, carbon tetrachloride, toluene, and o-xylene were attributed to the lack of functional groups in these solvents that could interact with the gold nanoparticle surfaces. In contrast, the nonlinear results in polar solvents such as acetonitrile (CH_3CN), tetrahydrofuran (THF), 1,4-dioxane, dimethylformamide (DMF), and dimethyl sulfoxide (DMSO) could be due to the formation of complexes on the gold surface via direct charge-transfer interactions and electron injection.

The authors subsequently investigated the effect of the solvent chain length using alcohols with variable chain lengths, $CH_3(CH_2)_{x-1}OH$, where $x = 1$–10. As evident in Fig. 3.5b, as the chain length increases, the λ_{max} blueshifts display the opposite behaviour expected for changes in the refractive index. These results suggest that alcohols, as polar solvents, can interact with gold nanoparticles by donating nonbonding electrons from their –OH group to the gold surface. The blueshift caused by the increased chain length can be attributed to a high diffusion and interaction efficiency with gold nanoparticles for the alcohols with short chain lengths. The electron densities of the gold nanoparticle surfaces decreased after the nanoparticles interacted with the alcohols.

Ligands bound to the surfaces of gold nanoparticles can also influence the plasmon resonance band. Hence, ligand effects were determined via two types of surfactants: those with cationic groups, including cetyltrimethylammonium chloride ($C_{16}TAC$), $C_{10}TAC$, $C_{12}TAC$, and $C_{14}TAC$, and those with anionic groups, including sodium dodecyl sulphate (SDS), decylsodium sulphate (DSS), and sodium dodecyl benzene

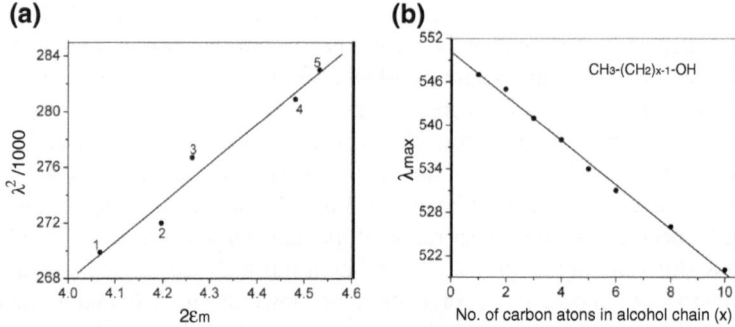

Fig. 3.5 **a** Plot of the square of the absorption maxima as a function of twice the medium dielectric function (ε_m was determined from the expression, $\varepsilon_m = n^2$). Integers *1*, *2*, *3*, *4*, and *5* on the curve represent cyclohexane, chloroform, carbon tetrachloride, toluene, and *o*-xylene, respectively. **b** Dependence of the observed peak position of the LSPR spectra of gold nanoparticles on the chain length of alcohol. Reprinted with permission from Ref. [9]. Copyright (2004) American Chemical Society

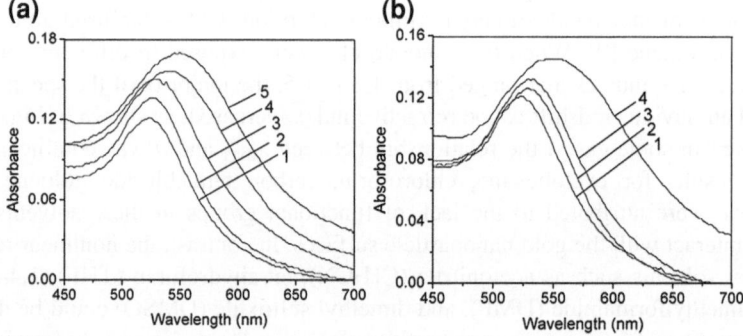

Fig. 3.6 **a** Absorption spectrum of gold nanoparticles (50 μM) (*1*) as prepared in THF and after modification with 0.1 mM of (*2*) C_{10}TAC, (*3*) C_{12}TAC, (*4*) C_{14}TAC, and (*5*) C_{16}TAC, respectively. **b** Absorption spectrum of gold nanoparticles (50 μM) (*1*) as prepared in THF and after modification with 0.1 mM of (*2*) DSS (*3*) SDS, and (*4*) SDBS, respectively. Reprinted with permission from Ref. [9]. Copyright (2004) American Chemical Society

sulphonate (SDBS). As shown in Fig. 3.6, as the chain length of the cationic surfactant increased, the plasmon resonance band gradually redshifted and also increased in intensity and linewidth. Similarly, the LSPR spectra also redshifted as the chain lengths of the anionic surfactants increased. The shells formed by the surfactants on the surfaces of the gold nanoparticles altered the refractive index surrounding the particle surface. Murray modified the Mie theory equation, taking into consideration the dielectric constant of the surfactant shell on the particle surface, as shown in Eq. (3.2) [10]:

$$\lambda^2 = \lambda_p^2[(\varepsilon^\infty + 2\varepsilon_m) + 2g(\varepsilon_s - \varepsilon_m)/3] \tag{3.2}$$

where λ_p is the bulk-metal plasmon wavelength, ε^∞ is the high-frequency die-lectric constant due to interband and core transitions, ε_m ($=n^2$) is the optical dielectric function of the medium, and ε_s is the optical dielectric function of the shell layer. When modified with surfactant, the gold nanoparticles were assumed to exhibit a core–shell structure, with the particle as the core and the surfactant forming the shell. Thus, g is the volume fraction of the shell layer expressed as below:

$$g = \left[(R_{core} + R_{shell})^3 - R_{core}^3\right] / (R_{core} + R_{shell})^3 \tag{3.3}$$

When the chain length of the surfactant increased, the volume fraction of the shell layer g also increased, inducing redshifts in the LSPR spectra. In addition, the die-lectric constant of the shell layer ε_s increased as the chain length became longer.

3.3   Effect of Substrate

The substrate plays an important role in the plasmon resonance of plasmonic nanoparticles [11]. Wax has proposed a theory to calculate the effects of the sub-strate and the environment on the scattered light by comparing the experimental and calculated scattering peak wavelengths of single silver nanoparticles [12]. A weighting factor α was introduced to facilitate the modelling of the changes in the surrounding refractive index based on Mie theory, as depicted in Eq. (3.4). In this work, two models were developed to calculate the weighting factor. For the first model, the related medium sensitive volume was presumed to be a shell with uni-form sensitivity that extends a one-particle-radius distance. The ratio between the medium volume above the substrate and the whole shell volume was the value of the weighting factor, 0.82. In the second model, the sensitivity of the plasmon-ics resonance to the surrounding medium was predicted to decrease exponen-tially from the particle surface to infinity. The weighting factor was calculated by integration along z and ρ in a cylindrical coordinate system, giving $\alpha = 0.7$. This method provided a calibration of the effect of the surrounding medium on the plas-mon resonance of the nanoparticles and increased the accuracy of the theoretical simulation of the LSPR band.

$$n_{eff} = \alpha \times n_{medium} + (1 - \alpha)n_{substrate} \tag{3.4}$$

Notably, when a substrate exhibits plasmon resonance, it will experience intense interactions with the nanoparticles near the substrate. Kik used single silver nanoparticles on a gold film to investigate the interactions between nan-oparticles and plasmonic substrates [13]. The plasmon resonance band of the silver nanoparticles was tuned from blue wavelengths to the near-IR region by the coupling of the nanoparticles and the gold film. Figure 3.7c shows a dark-field image of single silver nanoparticles on a glass slide. The majority of the silver nanoparticles exhibit blue or blue-green colours, with scattering peak

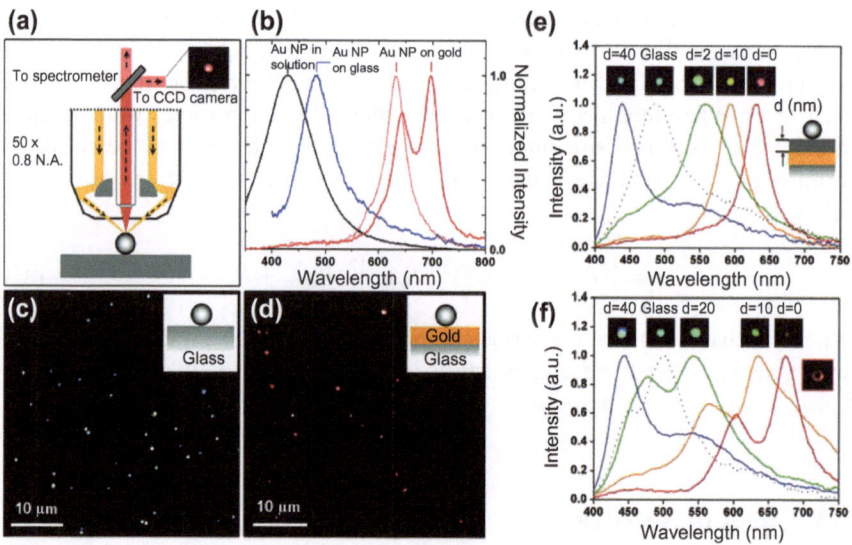

Fig. 3.7 **a** Schematic of the experimental arrangement for single-particle dark-field microscopy and spectroscopy using a reflected light dark-field objective. **b** *Black line* UV–Vis extinction spectrum of an aqueous solution of spherical silver nanoparticles with an average diameter of 60 ± 5 nm used in the experiment; *blue* and *red line*: examples of optical scattering spectra for single silver nanoparticles deposited on a glass substrate (*blue line*) and on a 50-nm gold film (*red solid line* and *dotted line*). **c** and **d** True colour dark-field microscopy images of isolated silver nanoparticles on a plain glass substrate (**c**) and on a 50-nm gold film deposited on a glass substrate (**d**). The *insets* in (**c**) and (**d**) schematically show the cross section of the corresponding substrates. **e–f** Examples of optical scattering spectra for silver nanoparticles on different substrates exhibiting a single resonance peak (**e**) and exhibiting two resonance peaks (**f**). The *inset* in **e** schematically shows the cross section of the substrate with a silica spacer layer of varying thickness d (nm) on top of a 50-nm gold film supported by a glass substrate. Note that the cross sections of the substrates are the same for (**e**) and (**f**), but the spectra were taken from different individual nanoparticles. Above each spectrum in (**e**) and (**f**) is the corresponding true colour dark-field image of the silver nanoparticle. The *dotted lines* represent single-particle spectra of silver nanoparticles on plain glass substrate. The highlighted rightmost dark-field image in (**f**) is an example of a vertically polarised resonance of a silver particle directly on a gold film, leading to a donut-shaped optical image. Adapted with permission from Ref. [13]. Copyright (2010) American Chemical Society

wavelengths of 430–490 nm. When the silver nanoparticles were deposited onto a 50-nm gold film on a glass substrate, their colours changed to red or orange-red, as shown in Fig. 3.7b and d, thereby confirming the coupling between the nanoparticles and the gold film. The investigators also observed that the spectra of some of the nanoparticles contained one predominant peak with a weaker shoulder, whereas the spectra of other nanoparticles contained two clearly separated scattering peaks. Then, the authors investigated the distance between the silver nanoparticles and the gold film by fabricating substrates with various silica spacer layer thicknesses, d, as shown in Fig. 3.7e. For $d = 40$ nm, the peaks in the scattering spectra of the silver nanoparticles blueshifted compared with the peaks of nanoparticles deposited on the glass substrate. When d was less than 20 nm, the coupling between the particles and the gold film induced an obvious redshift, as depicted in Fig. 3.7e and f. To explain this phenomenon, a dipole–dipole interaction model was applied to qualitatively expound the plasmon resonance shifts of the nanoparticles deposited on gold films. Two polarised modes, including horizontal and vertical surface plasmon modes, were presented. For the thin spacer layer ($d < 20$ nm), in the case of the particles with two peaks, the vertical mode-induced electric field was twice as strong as that induced by the horizontal mode. The longer wavelength peaks of the particles with smaller spacer layer thicknesses were attributed to the vertical mode, whereas the shorter wavelength peaks were due to the horizontal mode. When d was greater than 60 nm, the redshifts in the scattering spectra of the single nanoparticles were mainly due to the polarisation of silica glass. For 20 nm $< d <$ 60 nm, both the polarisability of the silica film and the charge response in the metal film affected the particle plasmon resonance frequency. In addition, Nordlander and co-workers used the plasmon hybridisation concept to theoretically examine the plasmonic properties of a sliver nanocube on a dielectric substrate [14].

3.4 Refractive Index-based Biosensors

Based on the effect of the surrounding medium on plasmonic nanoparticles, a novel sensor platform was fabricated using membrane-coated nanoparticles for the detection of streptavidin binding to biotinylated lipids, as shown in Fig. 3.8 [15]. The antigen–antibody binding event induced shifts in the plasmonic resonance bands, indicating the occurrence of reactions on and within the biomembranes. Moreover, the membrane on the surface of the nanoparticles provided an active area for the interactions between proteins, which enhanced the specificity of the system. Notably, a new method of fast single-particle spectroscopy was developed to monitor the spectral shifts of many nanoparticles in parallel. This method improved the efficiency of the collection of scattering spectra and enabled the analysis of complex processes on the basis of equilibrium coverage fluctuations.

Moving beyond the detection of large biomolecules, Song et al. [16] developed a sensor for the detection of mercuric ions; the sensor exhibited a detection range of 10 μM–100 pM and high selectivity based on the sensitivity of plasmonic gold nanoparticles to their surrounding medium . Chilkoti and co-workers detected protein binding on a single gold nanorod using plasmon resonance scattering spectroscopy. The binding events induced changes in the surrounding medium, and the sensitivity of the nanorods was estimated from a linear fit with a slope of 262 nm/RIU [17].

Compared with single-nanoparticle biosensors, core–shell multiarray LSPR-based nanochips provided convenient and low-cost diagnoses with high sensitivity

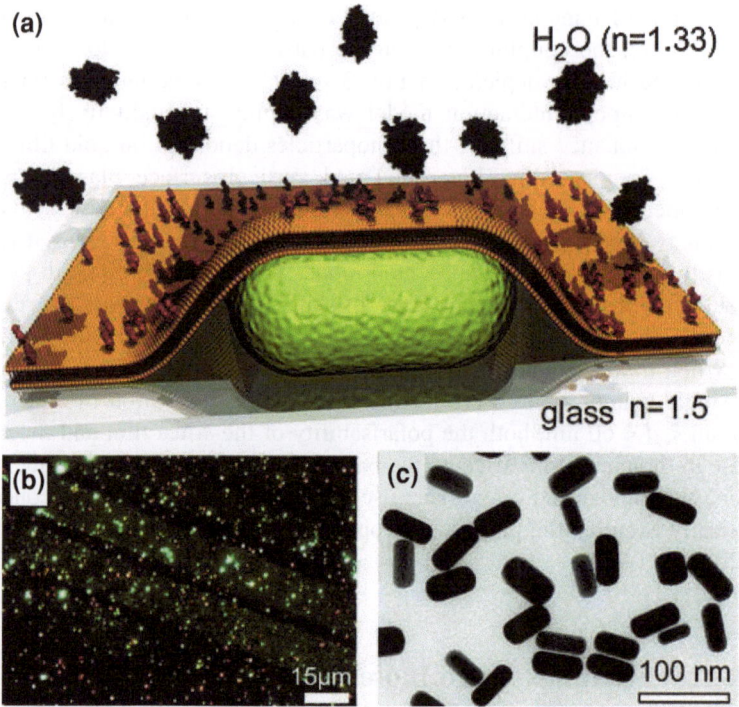

Fig. 3.8 **a** Schematics of a gold nanorod (*yellow*) coated with a partly biotinylated (*red*) membrane (*orange*) and exposed to streptavidin (*green*). The scheme presents the ideal case of a complete membrane coverage without any defects. The glass support has a refractive index of $n = 1.5$; the aqueous buffer has a refractive index of $n = 1.33$. **b** Dark-field image of gold nanorods (*bright spots*) covered by a lipid membrane. The membrane (*greenish background*) is structured on this substrate by a micromolding in capillaries. **c** Transmission electron microscope (*TEM*) image of gold nanorods used in this work (mean length 56 ± 5 nm, width 26 ± 5 nm, aspect ratio 2.2 ± 0.4 nm, determined from 100 particles). Reprinted with permission from Ref. [15]. Copyright (2008) American Chemical Society

and selectivity, as shown in Fig. 3.9 [18]. Amino-group-modified 100-nm silica nanoparticles were used as the "core", and the top and bottom gold layers, which were deposited via thermal deposition, were used as the "shell". Approximately 300 nanoparticles were placed on the nanochip, and antibodies were immobilised onto the nanospots by self-assembly. The entire nanochip was less than 20 cm in length, as shown in Fig. 3.9. This microfabrication technology enabled the parallel detection of multiple target molecules using the LSPR absorption intensity, which was modulated by changes in the refractive index of the surrounding medium caused by the interactions of biomolecules.

To enhance the sensitivity of plasmonic nanoparticles, Van Duyne proposed three methods for decreasing the detection limits. First, large molecules, such as proteins and macromolecules, induce larger peak shifts than smaller molecules [19, 20].

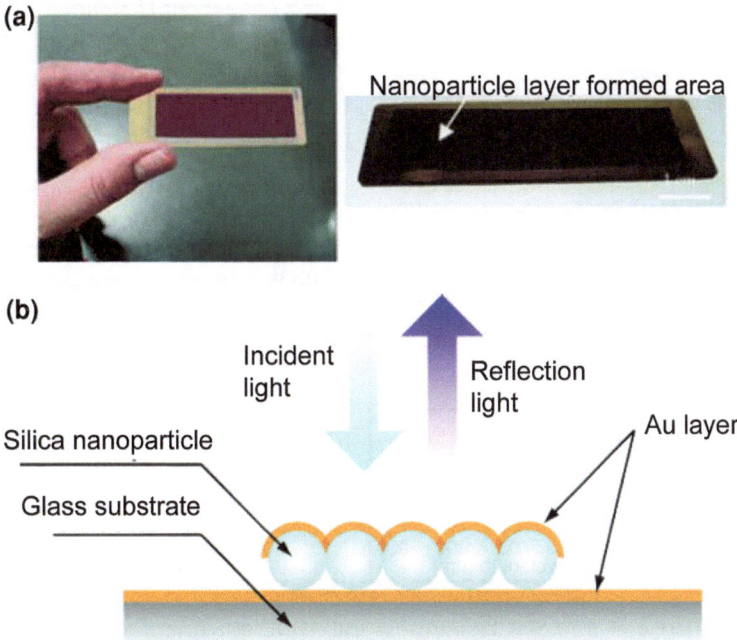

Fig. 3.9 a Photograph of the multiarray gold-capped nanoparticle layer substrate. The antibodies were immobilised onto the multiarray gold-capped nanoparticle layer substrate surface using a nanoliter dispensing system. **b** Construction of the multiarray LSPR-based nanochip. The surface-modified silica nanoparticles were aligned onto the gold-deposited glass substrate surface. Subsequently, the gold layer was deposited onto the silica nanoparticle layer. Adapted with permission from Ref. [18]. Copyright (2006) American Chemical Society

Thus, plasmonic nanoparticles were used to detect protein–protein interactions, such as biotin–streptavidin binding and interactions between amyloid β-derived diffusible ligands (ADDLs) and anti-ADDLs antibodies, as shown in Fig. 3.10 [21]. The binding of protein caused obvious LSPR peak shifts. Furthermore, the surface-confined binding constant was calculated as:

$$\Delta R = \Delta R_{\max} [K_{a, \text{surf}}[\text{anti-ADDL}]/(1 + K_{a, \text{surf}}[\text{anti-ADDL}])] \qquad (3.5)$$

where $\Delta R = \Delta\lambda_{\max}$, the LSPR peak shift for a given concentration; $\Delta R_{\max}$ is the maximum LSPR response; $K_{a,\text{surf}}$ is the surface-confined thermodynamic affinity constant; and [anti-ADDL] is the concentration of protein anti-ADDL. From this formula, $K_{a,\text{surf}}$ can be obtained for a fixed concentration of protein anti-ADDL.

The second signal enhancement method employs chromophores, which absorb visible light and match the LSPR spectra of nanoparticles. When chromophores adsorb onto nanoparticles, the overlap between the resonant bands of the chromophores and particles causes large spectral peak shifts that are enhanced threefold compared with nonresonant conditions, thereby enabling the sensitive detection

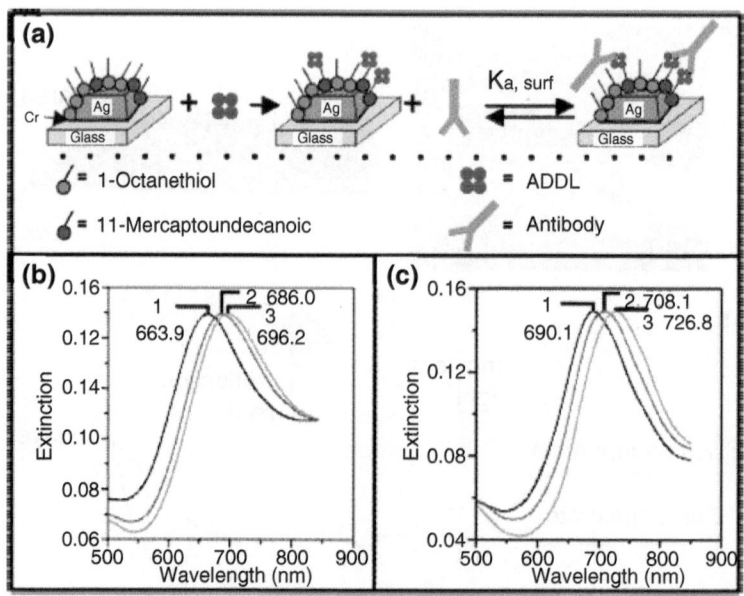

Fig. 3.10 Design of the LSPR biosensor for anti-ADDL detection. **a** Surface chemistry of the Ag nanoparticle sensor. Surface-confined Ag nanoparticles are synthesised using NSL. Nanoparticle adhesion to the glass substrate is promoted using a 0.4-nm Cr layer. The nanoparticles are incubated in a 3:1 1-OT/11-MUA solution to form a SAM. Next, the samples are incubated in 100 mM EDC/100 nM ADDL solution. Finally, incubating the ADDL-coated nanoparticles to varying concentrations of antibody completes an anti-ADDL immunoassay. **b** LSPR spectra for each step of the preparation of the Ag nanobiosensor at a low concentration of anti-ADDL antibody. Ag nanoparticles after modification with (**b–1**) 1 mM 3:1 1-OT/11-MUA, λ_{max} = 663.9 nm (**b–2**) 100 nM ADDL, λ_{max} = 686.0 nm, and (**b–3**) 50 nM anti-ADDL, λ_{max} = 696.2 nm. All spectra were collected in a N_2 environment. **c** LSPR spectra for each step of the preparation of the Ag nanobiosensor at a high concentration of anti-ADDL. Ag nanoparticles after modification with (**c–1**) 1 mM 3:1 1-OT/11-MUA, λ_{max} = 690.1 nm (**c–2**) 100 nM ADDL, λ_{max} = 708.1 nm, and (**c–3**) 400 nM anti-ADDL, λ_{max} = 726.8 nm. All spectra were collected in a N_2 environment. Reprinted with permission from Ref. [21]. Copyright (2004) American Chemical Society

of small molecules binding to protein receptors [22, 23]. Therefore, this method was used to detect the interactions between a small molecule, camphor, and the cytochrome P450cam protein (CYP101), as shown in Fig. 3.11 [24]. CYP101 has an absorption band at approximately 417 nm, and after CYP101 binds with camphor, its absorption peak blueshifts to 391 nm. Figure 3.11 displays the LSPR shift before and after the addition of camphor for different silver nanoparticles with initial peak positions at approximately 636.1 and 421.4 nm. In the case of nanoparticles with a peak wavelength of 421.4 nm, the peak wavelength was redshifted after

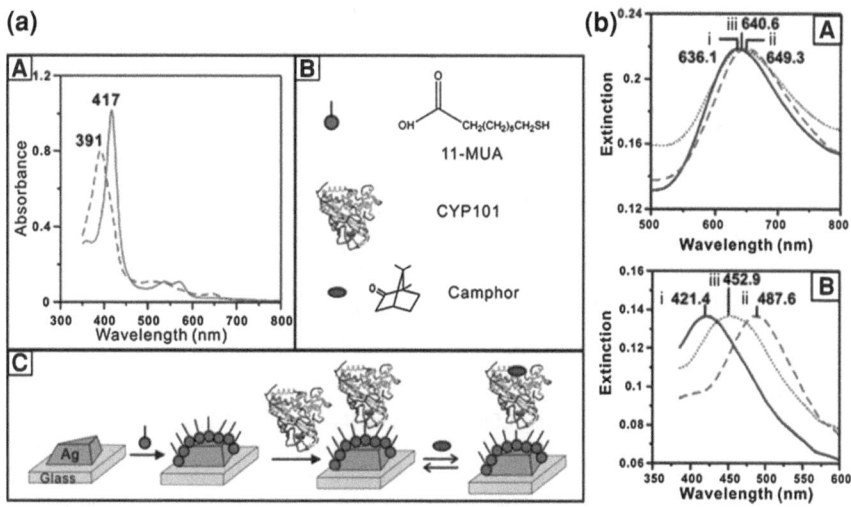

Fig. 3.11 **a** (*A*) UV–Vis absorption spectra of CYP101 (Fe^{3+}) (*green solid line*) with a Soret band at 417 nm (low spin) and camphor-bound CYP101-(Fe^{3+}) (*pink dashed line*) with a Soret band at 391 nm (high spin); (*B*) schematic notations of 11-MUA, CYP101, and camphor; (*C*) schematic representation of CYP101 protein immobilised Ag nanobiosensor, followed by binding of camphor. The Ag nanoparticles are fabricated using NSL (nanosphere lithography) on a glass substrate. **b** UV–Vis extinction spectra of each step in the surface modification of NSL-fabricated Ag nanoparticles and the wavelength-dependent LSPR shift plots. All extinction measurements were collected in a N_2 environment. A 200 μM camphor buffer solution was used: (*A*) a series of UV–Vis extinction spectra of Ag nanoparticles (i) $\lambda_{max,SAM} = 636.1$ nm (ii) $\lambda_{max,CYP101} = 649.3$ nm, and (iii) $\lambda_{max,CYP101-Cam} = 640.1$ nm; (*B*) a series of UV–Vis extinction spectra of Ag nanoparticles (i) $\lambda_{max,SAM} = 421.4$ nm (ii) $\lambda_{max,CYP101} = 487.6$ nm, and (iii) $\lambda_{max,CYP101-Cam} = 452.9$ nm. Reprinted with permission from Ref. [24]. Copyright (2006) American Chemical Society (color figure online)

the addition of CYP101 and was blueshifted in the presence of camphor. These nanoparticles were much more sensitive than those with a peak wavelength of 636.1 nm. The method was also applied to the detection of two different molecules that exhibit various influences on the chromophore absorption peaks.

The third enhancement method is based on the coupling of nanoparticles. When the distances between nanoparticles are less than 2.5 times the particle radius, their spectra undergo redshifts. Surface-confined silver nanoparticles were modified with biotin, and 20-nm solution-phase gold nanoparticles were conjugated to anti-biotin, as shown in Fig. 3.12 [25]. After the silver particles were treated with anti-biotin-modified gold nanoparticles, the spectra of silver particles showed a clear peak shift from 735.4 to 778.1 nm, with notable sensitivity.

These three methods can be used to enhance the sensitivity of plasmonic nanoparticles for the detection of molecules and provide guidance for the design of

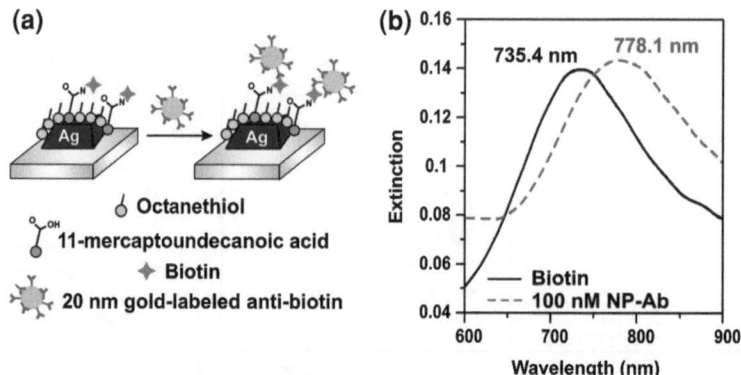

Fig. 3.12 Experiment schematic and LSPR spectra. **a** Biotin is covalently linked to the nanoparticle surface using EDC coupling agent, and antibiotin labelled gold nanoparticles are subsequently exposed to the surface. LSPR spectra are collected before and after each step. **b** LSPR spectra before (*solid black*) and after (*dashed red*) binding of antibiotin labelled nanoparticles, showing a $\Delta\lambda$ max of 42.7 nm. Adapted with permission from Ref. [25]. Copyright (2011) American Chemical Society (color figure online)

future biosensors. Researchers could employ appropriate approaches to discover unique applications. Furthermore, designing more efficient and creative biosensors based on the dielectric sensitivity of plasmonics to increase the sensitivity to single molecule level is still necessary and significant in the next few decades.

References

1. Mayer KM, Lee S, Liao H, Rostro BC, Fuentes A, Scully PT et al (2008) A label-free immunoassay based upon localized surface plasmon resonance of gold nanorods. ACS Nano 2:687–692
2. Nath N, Chilkoti A (2004) Label-free biosensing by surface plasmon resonance of nanoparticles on glass: optimization of nanoparticle size. Anal Chem 76:5370–5378
3. Otte MA, Sepúlveda B, Ni W, Juste JP, Liz-Marzán LM, Lechuga LM (2009) Identification of the optimal spectral region for plasmonic and nanoplasmonic sensing. ACS Nano 4:349–357
4. Zhao J, Zhang X, Yonzon CR, Haes AJ, Van Duyne RP (2006) Localized surface plasmon resonance biosensors. Nanomedicine 1:219–228
5. Mock JJ, Smith DR, Schultz S (2003) Local refractive index dependence of plasmon resonance spectra from individual nanoparticles. Nano Lett 3:485–491
6. McFarland AD, Van Duyne RP (2003) Single silver nanoparticles as real-time optical sensors with zeptomole sensitivity. Nano Lett 3:1057–1062
7. Anker JN, Hall WP, Lyandres O, Shah NC, Zhao J, Van Duyne RP (2008) Biosensing with plasmonic nanosensors. Nat Mater 7:442–453
8. Haes AJ, Zou SL, Schatz GC, Van Duyne RP (2004) Nanoscale optical biosensor: short range distance dependence of the localized surface plasmon resonance of noble metal nanoparticles. J Phys Chem B 108:6961–6968

9. Ghosh SK, Nath S, Kundu S, Esumi K, Pal T (2004) Solvent and ligand effects on the localized surface plasmon resonance (LSPR) of gold colloids. J Phys Chem B 108:13963–13971

10. Templeton AC, Pietron JJ, Murray RW, Mulvaney P (2000) Solvent refractive index and core charge influences on the surface plasmon absorbance of alkanethiolate monolayer-protected gold clusters. J Phys Chem B 104:564–570

11. Duval Malinsky M, Kelly KL, Schatz GC, Van Duyne RP (2001) Nanosphere lithography: effect of substrate on the localized surface plasmon resonance spectrum of silver nanoparticles. J Phys Chem B 105:2343–2350

12. Curry A, Nusz G, Chilkoti A, Wax A (2005) Substrate effect on refractive index dependence of plasmon resonance for individual silver nanoparticles observed using darkfield microspectroscopy. Opt Express 13:2668–2677

13. Hu M, Ghoshal A, Marquez M, Kik PG (2010) Single particle spectroscopy study of metal-film-induced tuning of silver nanoparticle plasmon resonances. J Phys Chem C 114:7509–7514

14. Zhang S, Bao K, Halas NJ, Xu H, Nordlander P (2011) Substrate-induced Fano resonances of a plasmonic nanocube: a route to increased-sensitivity localized surface plasmon resonance sensors revealed. Nano Lett 11:1657–1663

15. Baciu CL, Becker J, Janshoff A, Sönnichsen C (2008) Protein–membrane interaction probed by single plasmonic nanoparticles. Nano Lett 8:1724–1728

16. Song HD, Choi I, Yang YI, Hong S, Lee S, Kang T et al (2010) Picomolar selective detection of mercuric ion (Hg^{2+}) using a functionalized single plasmonic gold nanoparticle. Nanotechnology 21:145501–145507

17. Nusz GJ, Marinakos SM, Curry AC, Dahlin A, Höök F, Wax A et al (2008) Label-free plasmonic detection of biomolecular binding by a single gold nanorod. Anal Chem 80:984–989

18. Endo T, Kerman K, Nagatani N, Hiepa HM, Kim DK, Yonezawa Y et al (2006) Multiple label-free detection of antigen-antibody reaction using localized surface plasmon resonance-based core-shell structured nanoparticle layer nanochip. Anal Chem 78:6465–6475

19. Hall WP, Modica J, Anker J, Lin Y, Mrksich M, Van Duyne RP (2011) A conformation- and ion-sensitive plasmonic biosensor. Nano Lett 11:1098–1105

20. Haes AJ, Van Duyne RP (2002) A nanoscale optical biosensor: sensitivity and selectivity of an approach based on the localized surface plasmon resonance spectroscopy of triangular silver nanoparticles. J Am Chem Soc 124:10596–10604

21. Haes AJ, Hall WP, Chang L, Klein WL, Van Duyne RP (2004) A localized surface plasmon resonance biosensor: first steps toward an assay for Alzheimer's disease. Nano Lett 4:1029–1034

22. Ni W, Chen H, Su J, Sun Z, Wang J, Wu H (2010) Effects of dyes, gold nanocrystals, pH, and metal ions on plasmonic and molecular resonance coupling. J Am Chem Soc 132:4806–4814

23. Das A, Zhao J, Schatz GC, Sligar SG, Van Duyne RP (2009) Screening of type I and II drug binding to human cytochrome P450-3A4 in nanodiscs by localized surface plasmon resonance spectroscopy. Anal Chem 81:3754–3759

24. Zhao J, Das A, Zhang X, Schatz GC, Sligar SG, Van Duyne RP (2006) Resonance surface plasmon spectroscopy: low molecular weight substrate binding to cytochrome P450. J Am Chem Soc 128:11004–11005

25. Hall WP, Ngatia SN, Van Duyne RP (2011) LSPR biosensor signal enhancement using nanoparticle–antibody conjugates. J Phys Chem C 115:1410–1414

Chapter 4
Morphology- and Composition-Modulated Sensing

Abstract Nanoparticles with various sizes and shapes produce unique localized surface plasmon resonance bands and exhibit different physical and chemical properties. For instance, catalytic ability, sensitivity to changes in the surrounding medium, and biocompatibility are all dependent on the morphology of nanoparticles. In recent decades, various types of nanostructures have been fabricated to tune plasmon resonance bands, enhance the electromagnetic field around metal nanoparticles, and determine the relationship between the size and shape of nanoparticles and their LSPR band. In this chapter, we discuss the effect of morphology on plasmonic properties and the related applications.

Keywords Size of nanoparticles • Shape of nanoparticles • Composition of nanoparticles • Core–shell nanoparticles • Polarisation

4.1 Nanorods

LSPR property is dependent on the shape of nanoparticles that it is able to fabricate nanoplasmonics with different resonance band and sensing functions [1–3]. Particularly, nanorods have been widely applied in catalysis, biosensing, and photothermal therapy [4–7]. Nanorods with various aspect ratios have been fabricated, and their LSPR spectra can be predicted using the Gans theory [4]. Gans predicted that the surface plasmon mode can be divided into two parts for nanorods (also referred to as ellipsoids) when the dipole is constant. The Gans formula has been developed over the course of 40 years, and the polarisability of nanoparticles is now described by:

$$\chi_{x,y} = \frac{4\pi ab^2(\varepsilon_{Au} - \varepsilon_m)}{3\varepsilon_m + 3L_{x,y}(\varepsilon_{Au} - \varepsilon_m)} \tag{4.1}$$

Y.-T. Long and C. Jing, *Localized Surface Plasmon Resonance Based Nanobiosensors*, SpringerBriefs in Molecular Science, DOI: 10.1007/978-3-642-54795-9_4, © The Author(s) 2014

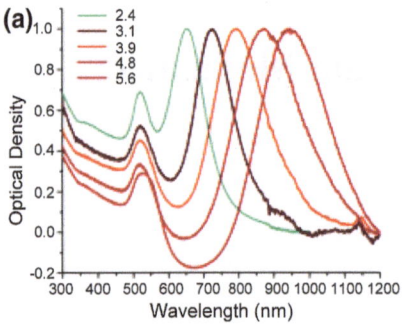

Fig. 4.1 a Surface plasmon absorption spectra of gold nanorods of different aspect ratios, showing the sensitivity of the strong longitudinal band to the aspect ratios of the nanorods. **b** TEM image of nanorods of aspect ratio of 3.9, the absorption spectrum of which is shown as the *orange curve* in panel (**a**). Reprinted with permission from Ref. [8]. Copyright (2006), American Chemical Society

where a and b denote the lengths of the nanoparticle along the x- and y-axes ($a > b$), ε_{Au} is the dielectric constant of Au, and ε_m is the dielectric constant of the medium. We can get $L_{x, y}$ as:

$$L_x = \frac{1 - e^2}{e^2}\left(-1 + \frac{1}{2e}\ln(\frac{1 + e}{1 - e})\right) \tag{4.2}$$

$$L_y = \frac{1 - L_x}{2} \tag{4.3}$$

where $L_{x, y}$ represents the depolarisation, e is the ellipticity, and the polarisability is easily related to C_{abs} and C_{sca} in Eq. (2.51).

Based on this theory, we conclude that, as the aspect ratio of nanorods increases, the LSPR band will shift to longer wavelengths, as shown in Fig. 4.1 [8]. By tuning the aspect ratio of nanorods, it is easy to obtain resonance band of nanoplamsonics at near-infrared region from 600 nm to more than 1,000 nm, providing multiple materials for diagnosis, theranostics, and mapping in vitro and in vivo.

4.2 Plasmonics Modulated by Different Morphologies

Over the past few decades, nanoparticles of various shapes and sizes have been designed and synthesised to investigate their optical properties and potential applications [9, 10]. Yang fabricated novel metallic nanostructures with shell-type plasmon geometries, such as gold nanograils, using colloidal lithography methods, as shown in Fig. 4.2 [11]. These nanograils contain three strongly coupled rings with sharp edges that significantly enhance the plasmon resonance of the nanostructures. In addition, these nanostructures were synthesised with different diameters to tune

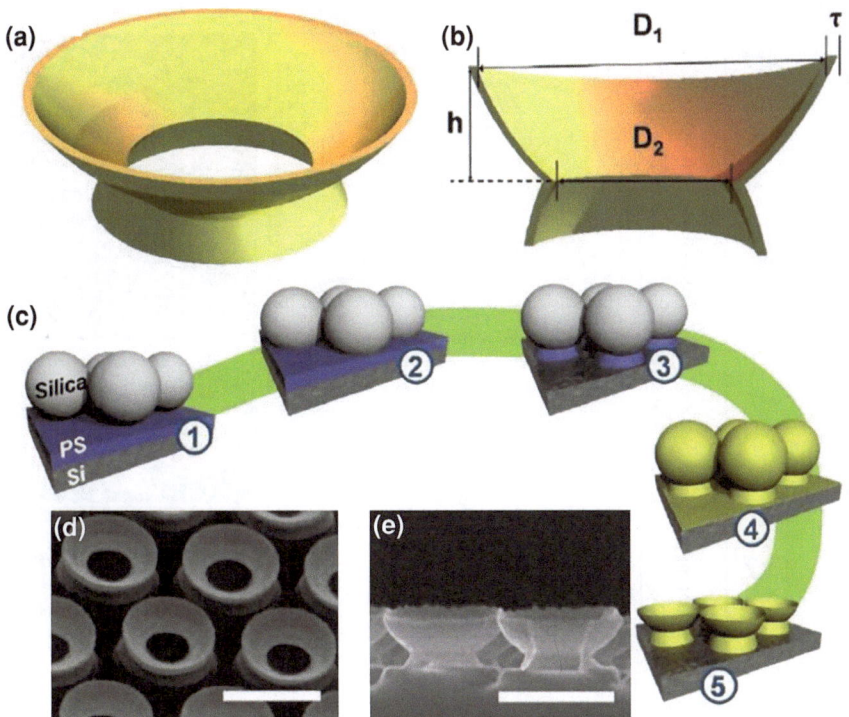

Fig. 4.2 Gold nanograil. **a** Gold nanograil with three sharp ring edges. **b** The gold nanograils are characterised by the overall thickness of the gold layer (τ), the diameter of the rings (D_1 and D_2), and the vertical separation between the rings (h). **c** Schematic representation of the gold nanograil fabrication: (*1*) silica colloidal particles (diameter 610 nm) are spin-coated onto the 150-nm-thick polystyrene film, (*2*) the particles are embedded in the polystyrene film, (*3*) a two-step reactive ion etching process with CF_4 and O_2 is performed to reduce the particle size and to remove the underlying polystyrene layer, (*4*) a thin layer of gold (40 nm) is sputtered onto the whole structure, and (*5*) Ar ion milling is performed to partly remove the deposited gold. The remaining silica particles are dissolved using a dilute HF solution (5 vol %). **d, e** SEM images of the gold nanograil arrays with tilted view angles of 30° and 90° (*side view*), respectively. Scale bars are 500 nm. Reprinted with permission from Ref. [11]. Copyright (2009), Wiley-VCH Verlag GmbH & Co. KGaA, Weinheim

the LSPR band in the range of 600–2,000 nm. For nanograils with $D_1 = 471$ nm, $D_2 = 225$ nm, $\tau = 15$ nm, and $h = 118$ nm, their reflectance spectrum shows two pronounced dips at 736 and 1754 nm. The spectrum is similar to the simulations by finite difference time domain (FDTD), as shown in Fig. 4.2. These nanostructures, which have a bulk refractive index sensitivity of 1,403 nm/RIU because of their large surface area, are far more sensitive to changes in the surrounding medium.

A new hybrid core–shell nanorod shaped like rice, termed "nanorice", was designed by Halas [12]. Nanorice consists of spindle-shaped Fe_2O_3 cores coated with Au shells of various thicknesses, as shown in Fig. 4.3a. Small gold

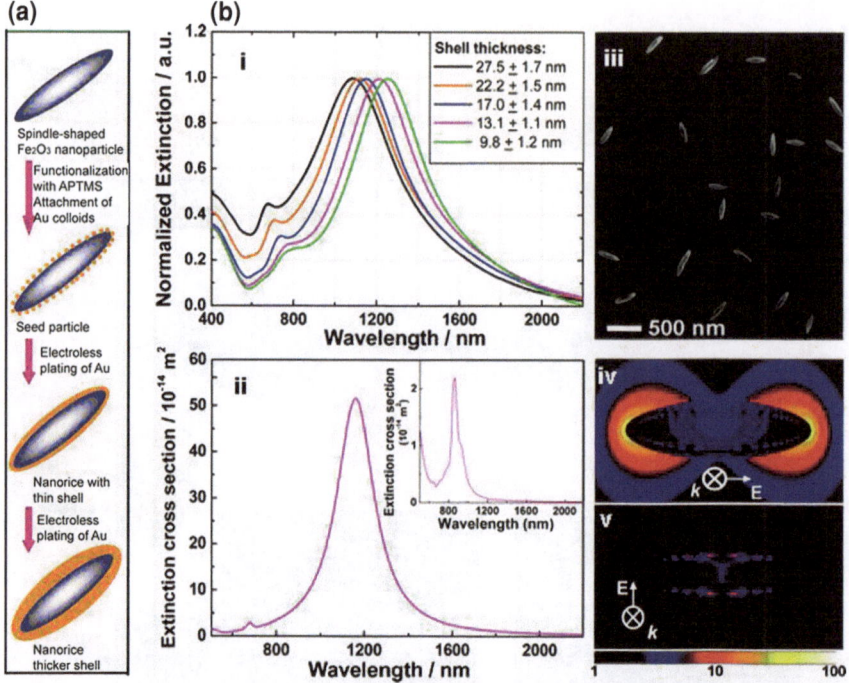

Fig. 4.3 **a** Schematics of the fabrication of haematite–Au core–shell nanorice particles. **b** (*i*) Extinction spectra of haematite–Au core–shell nanorice with different shell thicknesses. Two plasmon peaks are observed for each sample. The plasmons at longer and shorter wavelengths are the longitudinal and transverse plasmons, respectively. The samples measured are monolayers of isolated nanoshells immobilised on PVP-glass slides. (*ii*) Calculated far-field extinction spectra of the nanorice with incident polarisation along the longitudinal and (inset) transverse axis of a nanorice particle using FDTD. (*iii*) A SEM image of a monolayer of nanorice particles (shell thickness of 13.1 ± 1.1 nm) on a PVP-glass slide. The nanorice particle employed for the FDTD simulations is composed of a haematite core with longitudinal diameter of 340 nm and transverse diameter of 54 nm surrounded by a 13-nm-thick Au shell. Near-field profile of the nanorice under resonance excitations: (*iv*) incident polarisation along the longitudinal axis, $\lambda_{ex} = 1{,}160$ nm, and (*v*) incident polarisation along the transverse axis, $\lambda_{ex} = 860$ nm. Adapted with permission from Ref. [12]. Copyright (2006), American Chemical Society

nanoparticles (ca. 2 nm) are first immobilised on the surface of (3-aminopropyl)trimethoxysilane (APTMS)-functionalised Fe_2O_3 particles, which provide nucleation sites for the formation of gold shells. The extinction spectra of the nanorice with various shell thicknesses are displayed in Fig. 4.3b. As the shell thickness increases, the extinction spectra exhibited clear blueshifts. A simulation of the far-field extinction spectra of the nanorice using FDTD analysis revealed that the transverse plasmon mode was much weaker than the longitudinal mode. These results indicated that the surface asperities of the nanostructures enhanced the LSPR intensity substantially (>7,000 for the specific nanorice geometry). In addition, the longitudinal plasmon resonance band of the nanorice was highly sensitive to

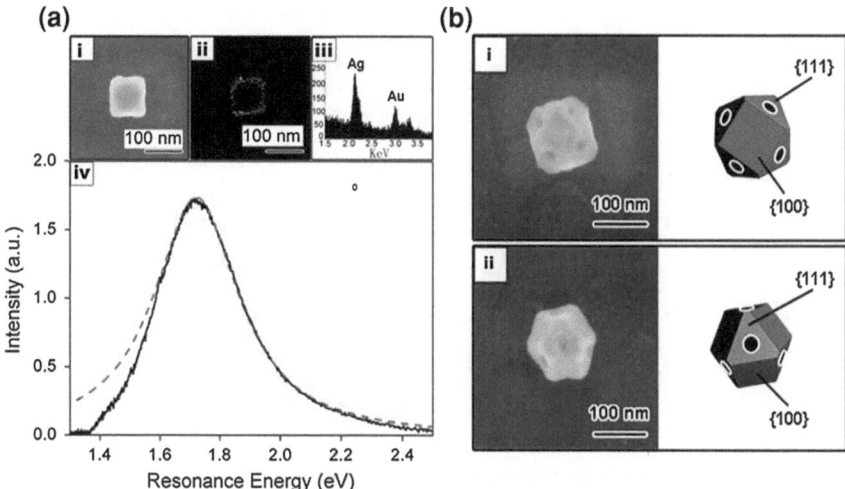

Fig. 4.4 **a** (*i*) Secondary electron SEM image of a Au–Ag nanobox. (*ii*) Backscattering SEM image of the nanobox. The wall thickness can be determined from this image. (*iii*) EDAX data for the nanobox, giving a Au:Ag ratio of 1:2. (*iv*) Optical scattering spectrum recorded using dark-field microscopy. The dashed line shows a Lorentzian fit to the data. **b** Different orientations of the nanocages on the substrate. *i* A nanocage with a {100} surface contacting the substrate (type I). (*ii*) A nanocage with a {111} surface contacting the substrate (type II). Adapted with permission from Ref. [17]. Copyright (2007), American Chemical Society

the surrounding refractive index, with a sensitivity of 801 nm/RIU, far exceeding the transverse plasmon mode with a sensitivity of 103 nm/RIU. This new nanostructure geometry can provide an attractive material for LSPR sensing and SERS detection because of its strong plasmon resonance band and high sensitivity.

Hollow nanoparticles, Au–Ag nanoboxes, and nanocages have been developed and modified on gold-coated ITO substrates [13–16]. The scattering spectra of these single nanostructures were collected to illustrate their LSPR properties, as shown in Fig. 4.4 [17]. In the case of the nanoboxes, their corners were truncated to give {111} facets with holes. Either the {100} facets or the {111} facets of the nanocages were in contact with the substrate, and each morphology produced unique LSPR bands. The scattering spectra of the nanocomposites with edge lengths of 80–160 nm exhibited plasmon resonance peaks in the range of 1.5–1.8 eV with broad full width at half-maximum (FWHM) values due to the combination of electron-surface scattering and radiation damping effects. Moreover, the nanocomposites exhibited a high sensitivity of ca. 360 nm/RIU to the surrounding refractive index.

Because of the importance of particle size and shape, many methods of fabricating nanoparticles have been developed [18–21]. Long proposed a template to prepare nanoparticles in a channel protein, stable protein 1 (SP1, ring diameter of 11 nm, inner pore of 2–3 nm, and width of 4–5 nm), using electrochemistry and monitoring the growth process under dark-field microscopy, as shown in Fig. 4.5 [22].

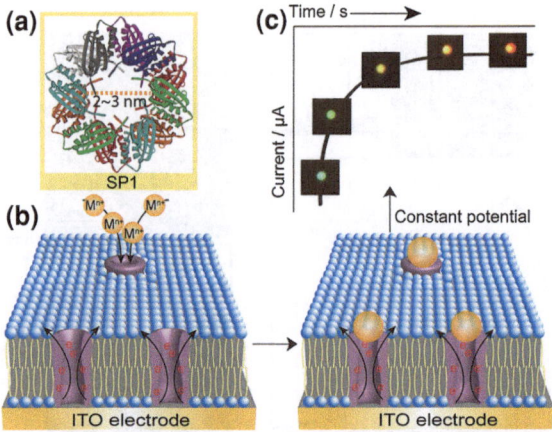

Fig. 4.5 Electrodeposition in the SP1 generated nanochannels, and Ag, Au, and Cu NPs deposited on the SP1-HBM/ITO template using a developing solution of AgClO$_4$, HAuCl$_4$, and CuSO$_4$ with the standard deposition potentials of -0.05, -0.05, and -0.6 V versus Ag/AgCl, respectively. **a** Structure of SP1. **b** SP1-HBM/ITO template. **c** The images of typical colour changes of a single Ag NP changing from *blue* to *red*, as the electrodeposition time is increased, indicating the in situ and real-time monitoring of the growth process of single NPs on the SP1-HBM/ITO template. Reprinted with permission from Ref. [22]. Copyright (2012), WILEY-VCH Verlag GmbH & Co. KGaA, Weinheim

An ITO slide was modified using hybrid bilayer membranes (HBMs), which are nonconductive. After the SP1 self-assembled on the HBMs, electrons and ions could transfer between the solutions and the ITO electrode through the SP1 channel. Thus, the nanoparticles grew in response to an applied reduction potential in the presence of metal ions in solution. This method provides an approach to fabricate nanoparticle arrays, and the density of the nanoparticles can be modulated through the concentration of SP1.

As single-nanoparticle detection has attracted more and more interest, analysing the mass of single nanoparticles has become critical. Knowledge of nanoparticle mass can eliminate the average effect of the bulk solution and prevent random events for single nanoparticles. Therefore, Long [23] developed a novel method to investigate the wavelength change and size distribution of numerous nanoparticles using the RGB (red, green, and blue) information from dark-field images. In this work, after obtaining a dark-field image of many single nanoparticles, the researchers assigned RGB values to every colour spot and then used these values to calculate the spectral peak wavelengths and diameters of the particles. This method provides a simple way to obtain statistical data for thousands of nanoparticles within several minutes using a common laptop computer. As shown in Fig. 4.6, the wavelengths and diameters of 1,766 particles were calculated in 3 min. Furthermore, this method can also be applied in cell imaging to overcome strong light scattering of tissues. Therefore, the scattering light of nanoparticles in cells could be easily captured using this RGB-based approach.

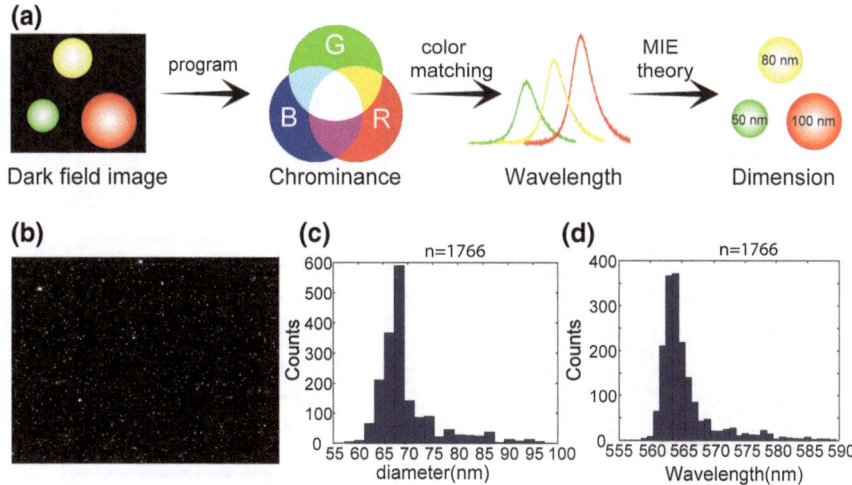

Fig. 4.6   a Calculation process of the RGB-based method. **b** Dark-field image and **c** calculated wavelength peaks and **d** diameter distribution of 1,766 GNPs. Adapted with permission from Ref. [23]. Copyright (2012), American Chemical Society

A new method for studying the 3-dimensional morphology and corresponding LSPR scattering spectra of single silver nanoparticles through a combination of atomic force microscopy (AFM) and dark-field microscopy was developed by Xu and co-workers [24]. Nanoparticle arrays were fabricated in microwindows on a glass slide through photolithography, which allowed the 3D morphology, dark-field images, and scattering spectra of single silver nanoparticles with different shapes to be monitored, as shown in Fig. 4.7. Five distinct nanoparticle shapes, including triangles, trapezoids, circles, hexagons, and parallelograms, were observed in one microwindow and were found to exhibit peak wavelengths of 506, 537, 548, 603, and 629 nm with a small shoulder-peak wavelength of 470 nm, respectively. As the shape of the nanoparticles transitioned from spheres to particles with sharper tips, the scattering spectra gradually shifted to longer wavelengths. These scattering spectra exhibited good agreement with the digital differential analyzer (DDA) theoretical calculations.

Schultz confirmed that the scattering spectra of nanoparticles with sharper tips have lower energies and longer peak wavelengths by comparing spherical, pentagonal, and triangular silver nanoparticles, as shown in Fig. 4.8 [25]. The scattering spectra of the single triangular particles were redshifted by more than 200 nm compared with peaks in the spectra of the spheres and by 100 nm compared with the peaks in the spectra of rounded nanotriangles.

Van Duyne [26] has proposed that both the structure-dependent frequency and the linewidth in an LSPR spectrum are important factors in sensing applications. To elucidate the influence of the structure of nanoparticles, including their size and shape, on the LSPR spectra, experimental data from individual bipyramidal silver

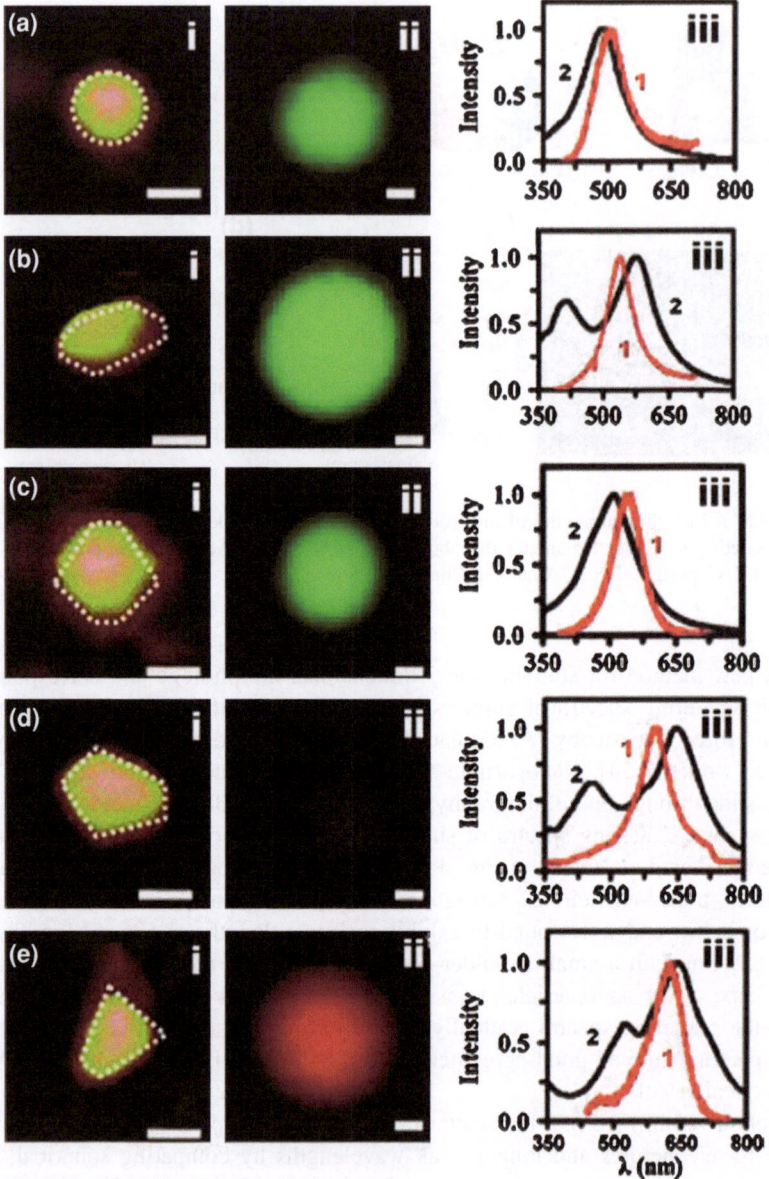

Fig. 4.7 Characterisation of shape-dependent optical properties (LSPR spectra) of single Ag NPs determined using experimental measurements and theoretical calculations. **a–e** (*i*) AFM images, (*ii*) dark-field optical colour images, and (*iii*) normalised LSPR spectra: (*1*) experimental measurements and (*2*) theoretical calculations of single Ag NPs. All scale bars are 100 nm. Note that the scale bars in (*ii*) are used to measure the size of image arrays, but not the sizes of single NPs because NPs are imaged under the optical diffraction limit. Reprinted with permission from Ref. [24]. Copyright (2009), American Chemical Society

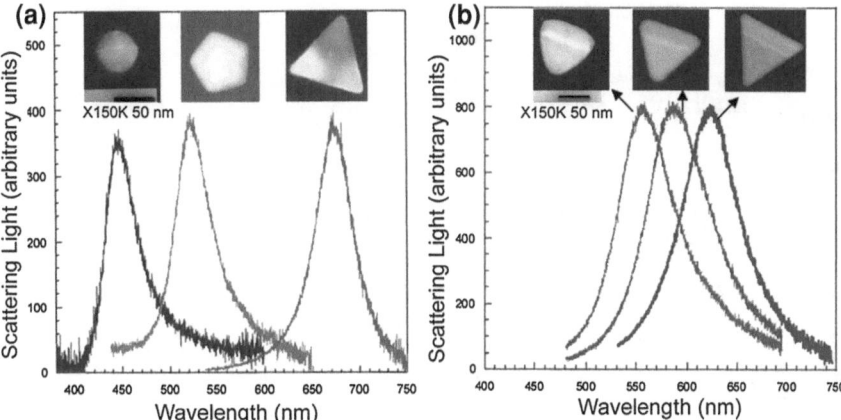

Fig. 4.8 **a** Typical optical spectroscopy measurements of individual silver nanoparticles. The figure shows the spectrum of an individual *red*, *green* and *blue particle*, and the high-resolution TEM images of the corresponding particle are shown above their respective spectrum. This example is a representative of the principle conclusion that the triangular-shaped particles appear mostly *red*, particles that form a pentagon appear *green*, and the *blue particles* are spherical. **b** Illustration of the particle shape and spectral modification by heating. The initial selected triangular silver nanoparticle had a spectral peak centred at 625 nm. After heating in air for 30 min at 200 °C, the particle shape changes and the triangle corners are more rounded, the spectral peak shifts to 585 nm, and the particle appears *orange*. An additional heating cycle for 20 min results in further rounding of the particle corners, and the particle appears *yellow/green* with the spectral peak centred at 555 nm. Reprinted with permission from Ref. [25]. Copyright (2002), AIP Publishing LLC

nanoparticles were collected. The bipyramid size was defined as the height of the equilateral triangle base, a. The truncation t was taken as the height of the triangle fitting in the empty corner of the equilateral base defined by the perfect bipyramid overlay. Typical scattering spectra of single bipyramids with different t/a ratios were obtained by dark-field microscopy. The peak at approximately 420 nm was attributed to the transverse mode of the nanoparticles that oscillates in the direction perpendicular to the base of the equilateral triangle, and the peak at approximately 600 nm was attributed to the longitudinal dipolar resonance. Based on the large number of scattering spectra of single nanoparticles, a two-parameter equation was applied to express the longitudinal plasmon resonance energy as a function of both size a and corner rounding t, as shown in Eq. (4.4). Both the size and shape of nanoparticles clearly affect their spectra. In addition, the authors confirmed that the plasmon frequency of sharp particles is more dependent on the size factor than that of rounded particles. Besides, the relationship between the plasmon linewidths of the bipyramids and their size and shape was taken into consideration. Size was observed to be the primary factor that influenced the spectral linewidth.

$$\text{LSPR}_{(ev)} = -0.0051(0.0003)a_{(nm)} + 0.019(0.003)t_{(nm)} + 2.69(0.06) \quad (4.4)$$

Van Duyne and co-workers also investigated the influence of size and shape on the wavelengths of scattering peaks and the FWHM of silver triangular nanoprisms

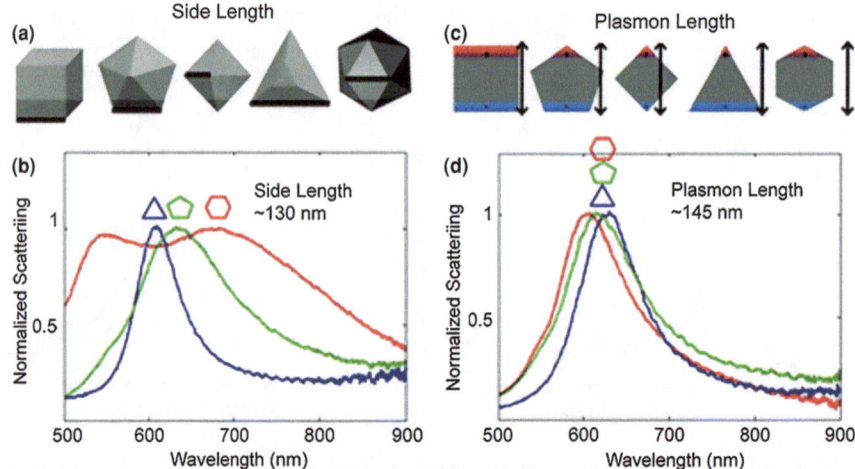

Fig. 4.9 Comparison between the side length and plasmon length as descriptors of nanoparticle size. **a** Definition of the side length for the shapes analysed. **c** Definition of plasmon length. **b**, **d** Representative single-particle spectra for Au triangles (*blue*), decahedra (*green*), and icosahedra (*red*) of similar side (**b**) or plasmon length (**d**). Reprinted with permission from Ref. [28]. Copyright (2012), American Chemical Society

(60–140 nm), including their edge length, roundness, truncation, and thickness [27]. As the aspect ratio increased, the LSPR band redshifted. In terms of truncation and roundness, when the tip truncation was reduced from 20 to 10 nm, a redshift occurred, and the nonradiative linewidth was also reduced. In contrast, further truncation reductions increased the FWHM due to shape-dependent radiative damping. Moreover, the researchers found that the experimental and calculated data were in good agreement when the electron collisions were parallel to the plane of the prism. In addition, nanoparticles with different sizes and truncations can exhibit the same LSPR peak wavelength or the same FWHM. The substrate factor only slightly affected the FWHM. Van Duyne also noted that other parameters, such as the surrounding medium, surface oxidation, and transverse tip truncation, can affect the LSPR properties of the particles. This study improves our understanding of LSPR and provides guidance for future LSPR applications.

A universal method to describe the effects of size on gold nanoparticles was developed for nanoparticles with fixed aspect ratios, well-formed vertices, and homogeneous small corner rounding, such as cubes, decahedra, icosahedra, octahedra, and truncated bitetrahedra, as shown in Fig. 4.9 [28]. Here, the size-dependent spectral bands and linewidths were correlated to a single parameter: the distance between the opposite-charge regions generated by the dipolar plasmonic electron oscillation, called the "plasmon length". This method used an intrinsic property of plasmonics to predict the plasmon resonance band and linewidth, thereby facilitating the calculation of particle properties as a function of shape.

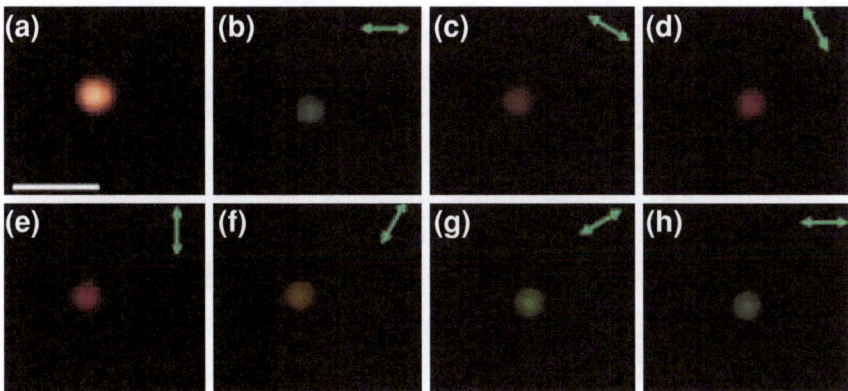

Fig. 4.10 Dark-field images of the single AuNR without polarisation (**a**) and at different polarisation angles (**b** 0°; **c** 30°; **d** 60°; **e** 90°; **f** 120°; **g** 150°; and **h** 180°). The scale bar represents 1 μm. *Green double arrows* represent the incident light polarisation. Reprinted with permission from Ref. [29]. Copyright (2011), Royal Society of Chemistry

4.3 Scattering Spectra Under Polarised Incident Light

Because of their anisotropy, metal nanoparticles such as nanorods exhibit polarisation-dependent plasmon resonance spectra. Kim investigated the scattering spectra of single gold nanorods with various aspect ratios at different polarisation angles [29]. They found that nanorods with diameters greater than 30 nm exhibited a variety of colours, including green, brown, red, and yellow, as shown in Fig. 4.10. The colours of the single nanorods were attributed to their inherent scattering spectra. When the incident light was polarised in parallel to the long axis of the nanorods, the longitudinal surface plasmon resonance band dominated, leading to a red colour. When the incident light was polarised parallel to the short axis, the transverse mode dominated, resulting in a green colour. At other polarisation angles, the colour of the nanorods reflected collaborative interactions between the longitudinal and transverse modes. Notably, nanorods less than 30 nm in diameter did not exhibit the rainbow-like colours because of the minute transverse contributions, which caused longitudinal dominance at the different angles.

In addition, a nanostar structure was prepared, as shown in Fig. 4.11 [30]. The scattering spectra of these complex, three-dimensional nanostars exhibit three peaks. Moreover, these multiple peaks are dependent on polarisation. Single-particle spectroscopy using dark-field microscopy was used to determine the scattering spectra of single nanostars. An analyser used to observe the polarisation of the scattering light of the nanostars was placed under the objective lens. As evident in Fig. 4.11, when the analyser was rotated, some of the peaks in the scattering spectra increased, whereas others decreased, depending on the analyser angle. Notably, the angle matched the edge of the nanostar tips. When the scattering spectra peaks overlapped considerably, they could not be separated by changing the

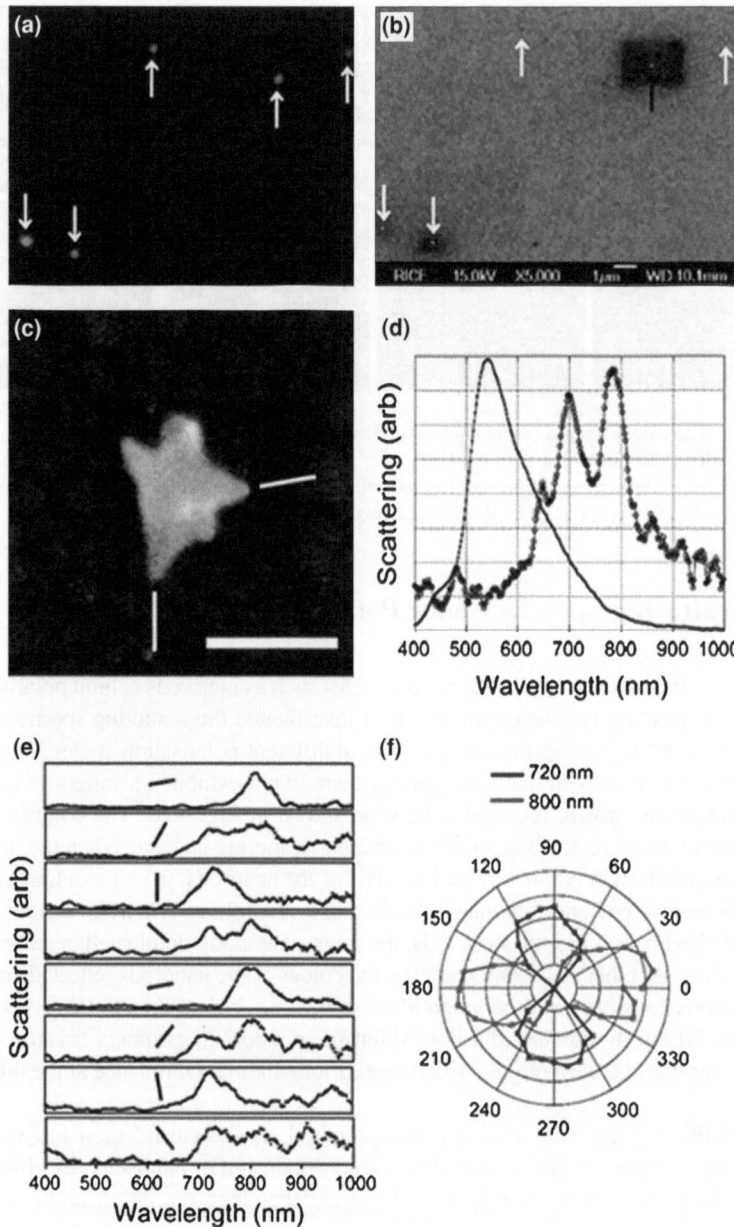

Fig. 4.11 Optical (**a**) and SEM (**b**) images of a field of gold nanostars demonstrate the correlation of their position relative to alignment marks (not shown). The *white arrow* indicates the
nanostar whose structure (**c** 100 nm scale bar) and scattering spectrum (**d** *open circles*) are displayed. The scattering spectrum of a 100-nm gold colloid is also plotted (**d** points). Eight spectra
at different analyser angles are plotted (**e**) as well as the peak heights for the 720 and 800 nm
peaks (**f**). Radial axis is arbitrary scattering. Reprinted with permission from Ref. [30]. Copyright
(2006), American Chemical Society

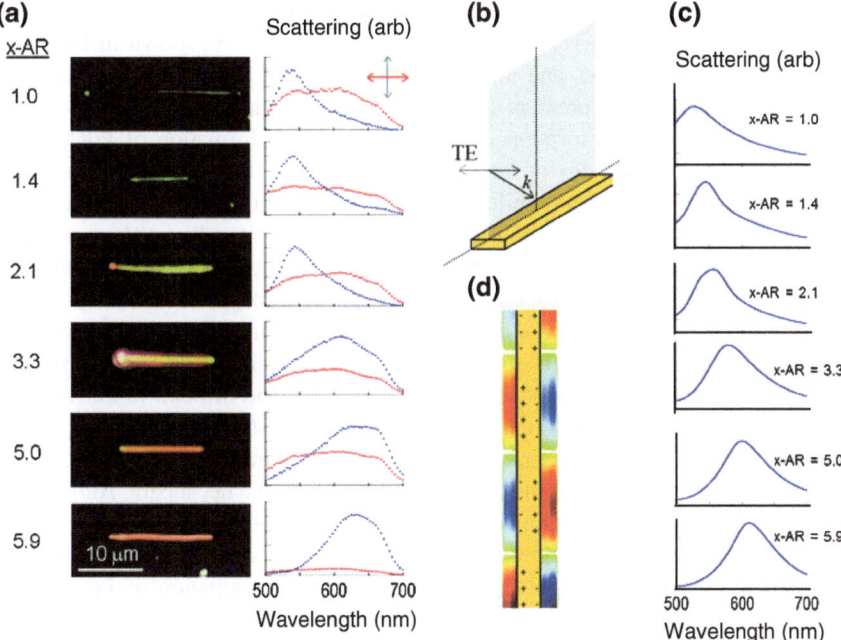

Fig. 4.12 **a** Dark-field micrographs (*left*) and corresponding single-nanobelt spectra (*right*). The given aspect ratios were determined by atomic force microscopy. The *blue spectra* are polarised transverse to the nanobelt, and the *red spectra* are polarised parallel to the nanobelt. **b** Gold nanobelt simulation geometry and **c** resulting spectra for cross-sectional aspect ratios that match (**a**). **d** The calculated charge distribution of the scattering mode. Reprinted with permission from Ref. [31]. Copyright (2011), American Chemical Society

angle. The nanostar structures also exhibited good sensitivity to the surrounding medium, with a sensitivity of ca. 600 nm/RIU. This three-dimensional structure could be used in further orientation studies that involve polarised light.

Hafner [31] fabricated gold nanobelts with cross-sectional dimensions less than 100 nm using an efficient chemical method. In this work, the authors found that the nanobelts exhibited sharp extinction spectra under transverse polarisation and broad peaks under the parallel mode. The transverse plasmon resonance band was dependent on the aspect ratio. These results were in good agreement with simulations by FDTD, as shown in Fig. 4.12.

4.4 Sensing Applications of Nanoparticles with Different Morphologies and Compositions

Variations in nanoparticle morphology can enable a wide variety of applications. Alivisatos investigated H_2 absorption and desorption on the surface of single Au/Pd core–shells with different shapes, facets, and Pd shell thicknesses via dark-field microscopy at room temperature (RT) [32]. The scattering spectral shifts indicated the uptake trajectories of H_2 and the reaction mechanism. As shown in Fig. 4.13, for single triangular plates and icosahedra, the reversible scattering spectral shifts were correlated with the pressure of H_2. The redshifts and blueshifts corresponded to the adsorption and desorption of H_2 because of the formation of PdH, which changed the refractive index of the particles' surfaces. The triangular plates exhibited an approximately 25-nm shift during the adsorption of H_2, which was much greater than the shifts of the icosahedra, which were less than 4 nm. This result was attributed to the high sensitivity of nanoparticles with high aspect ratios and sharp corners and edges. Notably, after several cycles of H_2 adsorption and desorption, the peaks in the scattering spectra of the nanoparticles were unable to return to their initial positions due to residual H atoms on the subsurface sites of Pd. The subsurface hydride and chemisorbed hydrogen were stable at RT and difficult to be removed. However, for the decahedral and hexagonal plate core–shell nanoparticles, an initial blueshift was detected during H_2 uptake, as shown in Fig. 4.13. There may be two possible explanations for the blueshift behaviour of the decahedra: Au/Pd interdiffusion and silicide formation. Upon H_2 absorption, the high pressure may have enhanced the interdiffusion of Au and Pd, effectively decreasing the gold particle size and leading to spectral blueshifts. Conversely, the formation of Pd_2Si could enhance the catalytic reaction and also resulted in blueshifts in the scattering. The triangular plates and icosahedra did not exhibit initial blueshifts, which may be due to weak interdiffusion and silicide formation. Approximately 60 % of the particles showed redshifts due to PdH formation, and 20 % exhibited an initial blueshift due to interdiffusion and silicide formation. The other 20 % showed no changes, which can be attributed to either the lack of a reaction or the simultaneous occurrence of both types of reactions. This work suggests that nanoparticles with different shapes and compositions have unique optical, catalytic, physical, and chemical properties.

4.5 Core–Shell Nanoparticles

Core–shell nanoparticles exhibit excellent properties, and their two or more component materials endow them with multiple functions, including catalytic abilities, biocompatibility, chemical reactivity, and magnetic properties. Furthermore, tuning the core–shell composition enables the modulation of the plasmon resonance band over a broad range, and extending their applications.

Gold/silica nanoshells have been investigated using high-resolution scanning electron and atomic force microscopy (SEM and AFM). The experimental plasmon resonance peak was in good agreement with values calculated via Mie theory [33]. Mirkin [34] reported a method for the synthesis of silica-encapsulated gold nanoprisms ($Au@SiO_2$) and monitored the coating process (Fig. 4.14a). Silica-coated plasmonic particles exhibit excellent stability, preventing etching in many chemical environments. Also, anisotropic silica growth was observed, with the edges of the prisms coated with less silica than the large triangular facets. Furthermore, the dielectric sensitivity of the fabricated nanoparticles was as high

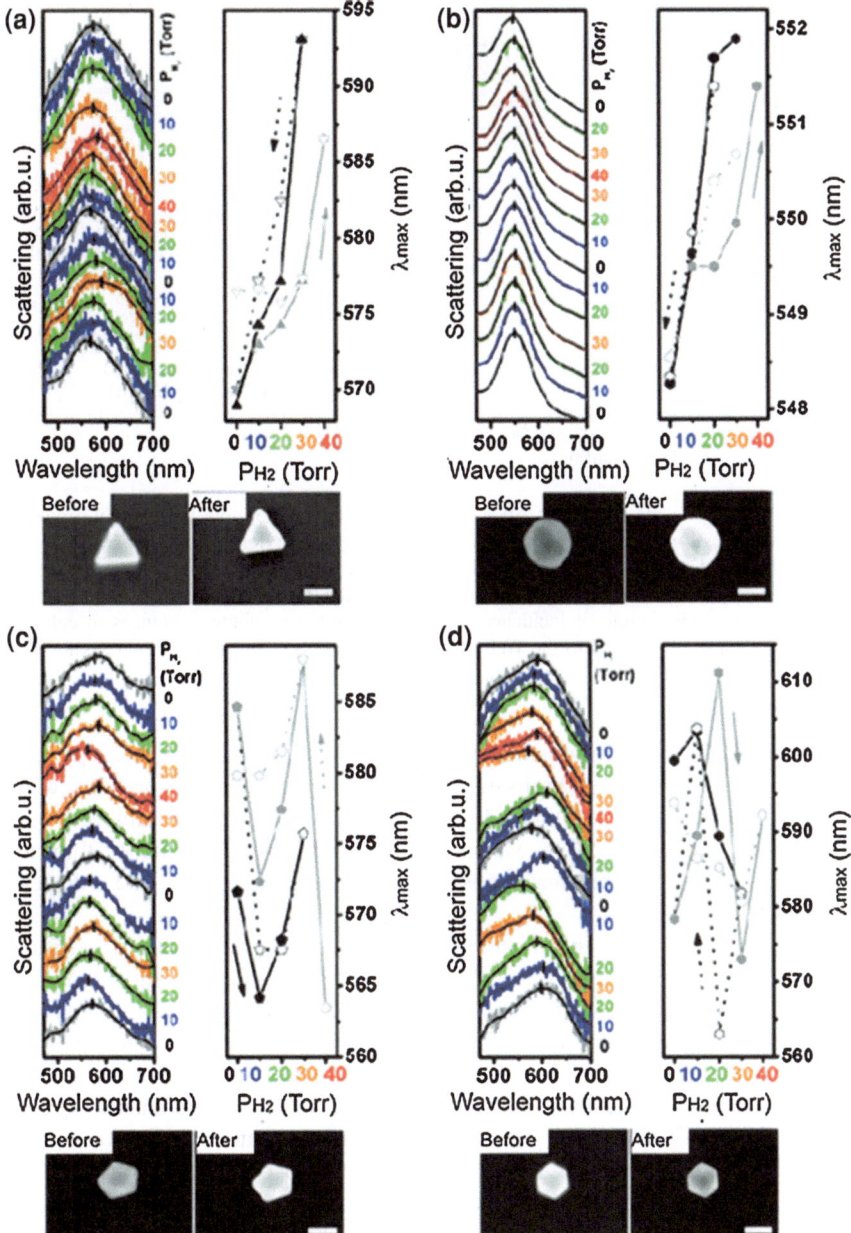

Fig. 4.13 Scattering spectra of a Au/Pd core/shell triangular plate (**a**), icosahedron (**b**), decahedron (**c**) and a hexagonal plate (**d**). Two cycles of increasing and decreasing P_{H2} (first cycle, *black*; second cycle, *grey*) are shown. Absorption: solid triangles (**a**), solid hexagonss (**b**), solid pentagons (**c**), and solid hexagons (**d**). Desorption: open triangles (**a**), open hexagons (**b**), open pentagons (**c**), and open hexagons (**d**). The total *redshift* was ca. 25 nm for the triangular plate but <4 nm for the icosahedron. The spectral shift was <20 nm for the decahedron and >30 nm for the hexagonal plate. The scale bars in the SEM images represent 100 nm. All measurements were performed at RT. Adapted with permission from Ref. [32]. Copyright (2011), American Chemical Society

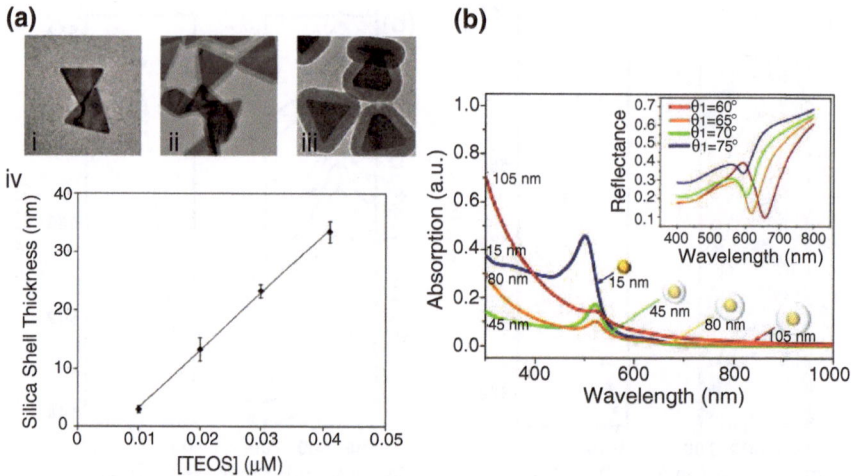

Fig. 4.14 **a** (*i–iii*) TEM images of Au nanoprisms with 3.0, 13.3, and 23.4-nm SiO$_2$ coating. (*iv*) Linear regression of silica thickness as a function of tetraethylorthosilicate (TEOS) concentration. Shell thickness (nm) = 984[TEOS] (μM) −6.60. **b** UV-Vis absorption spectra measured with GNPs and CSNPs of varying size. Inset: calculated SPR characteristics of a bare grating structure with varying angle of incidence (θ_i) (for interpretation of the references to colour in the text, the reader is referred to the Web version of the article). Adapted with permission from Ref. [34, 35]. Copyright (2010), American Chemical Society; 2012 Elsevier

as 737 nm/RIU. In addition, silica-coated gold nanoparticles have been prepared for the detection of DNA hybridisation localized by periodic linear gratings, which enhanced the optical signatures 36-fold compared with those achieved using conventional methods [35]. As shown in Fig. 4.14b, as the silica shell thickness increased, the plasmon resonance position showed a clear redshift and a decrease in intensity.

In addition to metal–silica structures, metal–metal core–shell nanoparticles have been studied for many years [36]. Gold/silver core–shells with various morphologies, such as cubes, truncated octahedra, octahedra, twinned hexagons and triangles, and five-twinned decahedra and nanorods, have been prepared using the seed-mediated method [37]. Another seed-mediated Au@Ag core–shell synthesis method is based on digestive ripening followed by annealing and enables the production of Au–Ag alloy NPs. This method provides quantitative control over the composition of the particles, allowing the modulation of the plasmon resonance band [38]. Because silver is more sensitive to dielectric changes, a tunable silver shell was deposited on gold nanoparticles using electrochemical methods. At a shell thickness of ~0.7 nm, the dielectric sensitivity was enhanced by 76 % compared with that of bare gold nanoparticles [39]. He [40] and co-workers used the Au@Ag core–shell structure to detect H$_2$S in living cells, as shown in Fig. 4.15. H$_2$S can react with Ag to produce Ag$_2$S in the presence of oxygen. Because the refractive index of Ag$_2$S (2.2) is much larger than that of Ag (~0.17), after reacting with H$_2$S, the Au@Ag core–shells showed a distinct redshift. Thus, incubation of the Au@Ag

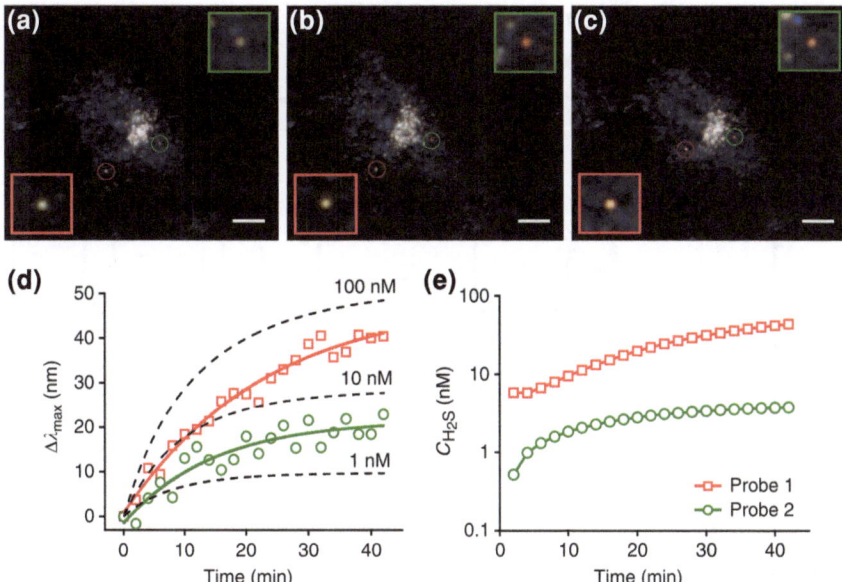

Fig. 4.15 Local variations of intracellular sulphide levels can be determined in real time. **a–c** Representative images showing the gradual colour changes of two individual PNPs after adding 0.1 μM Na$_2$S to the cell culture medium for **a** 2 min, **b** 26 min, and **c** 42 min. Scale bar, 10 μm. The *red* and *green square* inserts are enlarged images of the two circled PNPs. **d** Observed (*hollow dots*) and fitted (*lines*) time-dependent λ$_{max}$ shifts of the two particles. **e** Calculated time-dependent change in local sulphide concentrations surrounding the two particles according to the fitted results in d. Reprinted with permission from Ref. [40]. Copyright (2013), Nature Publishing Group

core–shells in a cell provides a sensitive approach to detect H$_2$S, with a linear logarithmic dependence on sulphide concentrations in the range of 0.01 nM to 10 μM.

Furthermore, novel composite nanoparticles were prepared, including fluorophore-doped silica cores and hemispherical gold shells separated by silica spacer layers [41]. Photoluminescence measurements indicated that the silica spacer decreased the quenching effect of the gold shells while maintaining the active functions of the core and shell materials. Because gold surfaces can be readily modified, this nanostructure could be a promising probe in biosensor applications, such as DNA hybridisation detection. Feldmann [42] constructed a Au@Au$_2$S core–shell structure for sensing at the single-nanoparticle level, as shown in Fig. 4.16. The shell thickness increased and the plasmon resonance band narrowed with the increase in the synthesis time. In particular, the dielectric sensitivity was much higher than that of Au NPs, which resulted in improved sensitivity of these plasmonic sensors. In addition, the formation process of Au$_2$S@Au structure was also investigated by increasing the Au shell thickness [43].

Combination of plasmonics and magnetic particles such as Fe$_3$O$_4$ endows the nanoparticles with excellent separation ability and promotes their applications in

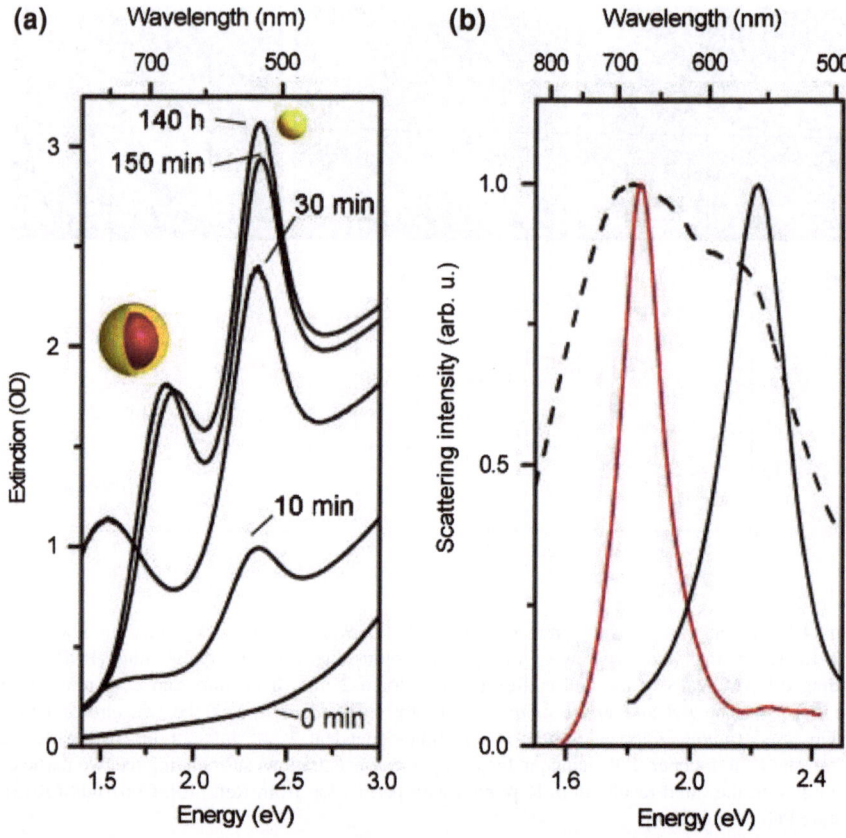

Fig. 4.16 **a** Successively recorded ensemble extinction spectra measured during the synthesis of Au$_2$S/Au nanoshells. The peak centred at 2.33 eV originates from the absorption of solid gold nanospheres with a diameter of ~5 nm. The second plasmon peak, shifting across the visible spectrum during the synthesis, arises from Au$_2$S/Au nanoshells. The given reaction times refer to the second addition of Na$_2$S (t = 0, 10, 30, 150 min, 140 h). **b** Scattering spectra of three individual nanoparticles: A Au$_2$S/Au nanoshell (*line in left*) and solid nanospheres with diameters of 40 nm (*solid line in right*) and 150 nm (*dashed black line*). Reprinted with permission from Ref. [42]. Copyright (2004), American Chemical Society

drug delivery and directional targeting [44, 45]. Sun [46] proposed a tunable Fe$_3$O$_4$/Au and Fe$_3$O$_4$/Au/Ag nanocore–shell fabrication method via the modulation of shell composition and thickness (Fig. 4.17). We can find that with the growth of Au shell, the nanoparticles showed redshift and the formation of Ag shell induced blueshift. This work provides a long-term modulation of resonance band of nanoparticles.

Because the formation of plasmonic core–shell structures can dramatically alter the LSPR band, Long fabricated biosensors to detect reduced nicotinamide adenine dinucleotide (NADH) [47]. NADH plays an important role in numerous biocatalysed processes, including energy metabolism, mitochondrial response, immunological function, ageing, and cell death. As a reductant, NADH can reduce

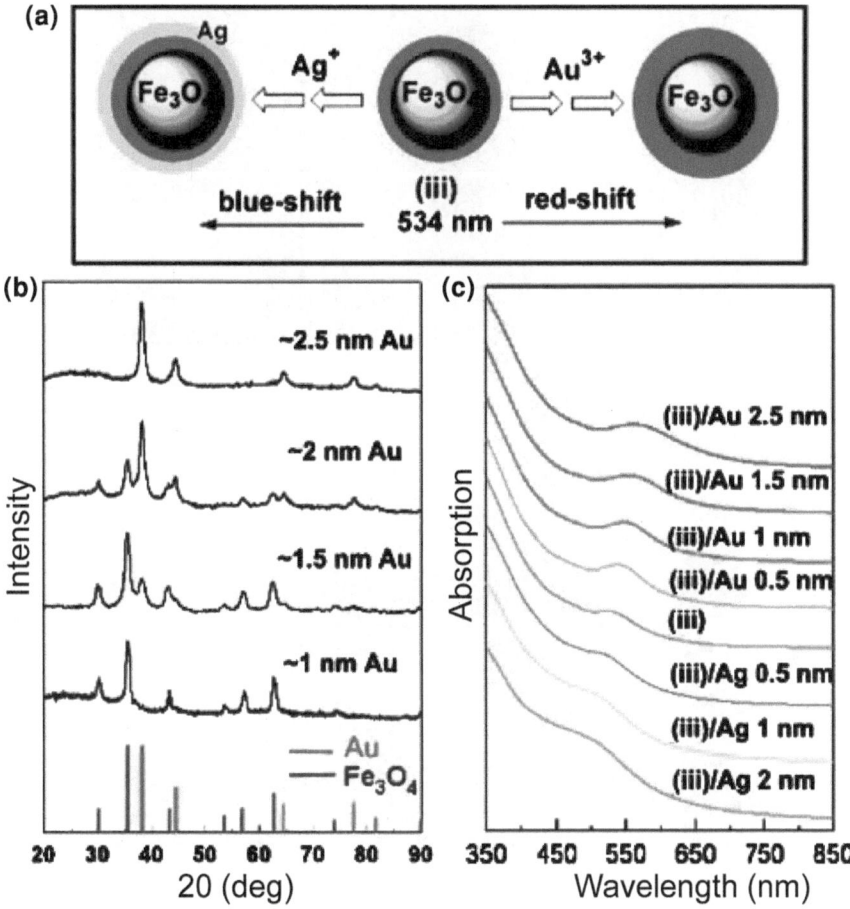

Fig. 4.17 a Schematic illustration of the formation of Fe_3O_4/Au and $Fe_3O_4/Au/Ag$ and the control on the plasmonic properties; **b** XRD of the Fe_3O_4/Au nanoparticles with various Au coating thickness; **c** UV-Vis absorption spectra of the core/shell Fe_3O_4/Au and $Fe_3O_4/Au/Ag$ nanoparticles with various Au and Ag coating thickness. Reprinted with permission from Ref. [46]. Copyright (2007), American Chemical Society

copper ions to metallic copper in the presence of gold nanoparticles as catalysts. The reduced copper atoms can then adsorb onto gold to form a Au–Cu core–shell structure, leading to redshifts in the plasmonic scattering spectra of the single nanoparticles. When HeLa cells are incubated with 50-nm gold nanoparticles for 24 h, the gold nanoparticles are clearly taken up by the HeLa cells and exhibit a green colour, as shown in Fig. 4.18f. The colour of the gold nanoparticles could be modulated by treatment with copper ions. In the presence of copper ions, the plasmonic nanoparticles gradually changed from green to red due to the reduction by NADH. Importantly, this method was used to monitor the metabolism of living cells and to screen the effects of anticancer drugs. In Fig. 4.18i, after the

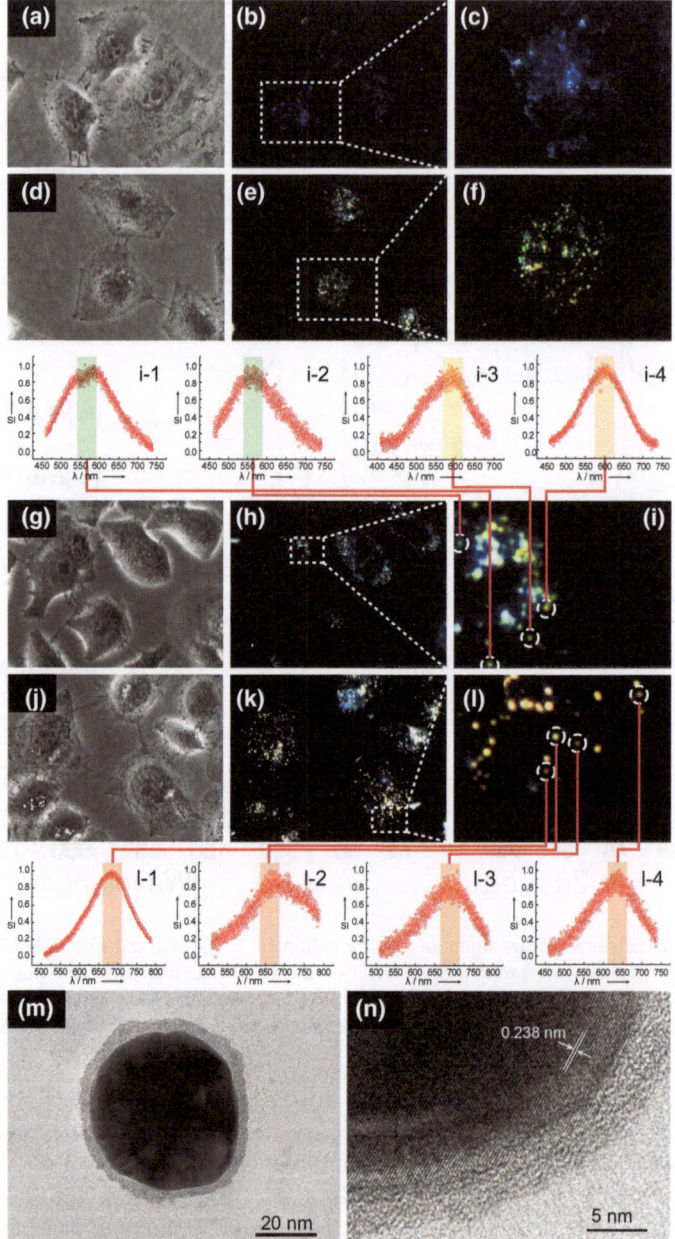

Fig. 4.18 a Bright-field images of HeLa cell. **b** DFM images of corresponding HeLa cell in (**a**). **c** The detail view of HeLa cell DFM images (**b**). **d** Bright-field images of HeLa cell after 24-h incubation with AuNPs. **e** DFM images of corresponding HeLa cell in (**d**). **f** The detail view of HeLa cell containing AuNPs DFM images (**e**). **g** Bright-field images of HeLa cell containing AuNPs with treatment by taxol (10 μm) and then incubation in TBS containing 50 μm $CuCl_2$ for 3 h. **h** DFM images of corresponding HeLa cell in (**g**). **i** The detail view of HeLa cell DFM images (**h**), *i-1* to *i-4*: corresponding scattering spectra of different AuNPs in living HeLa cell. **j** Bright-field images of HeLa cell containing AuNPs without treatment by taxol and then incubation in TBS containing 50 μm $CuCl_2$ for 3 h. **k** DFM images of corresponding HeLa cell in (**j**). **l** The detail view of HeLa cell DFM images (**k**), *l-1* to *l-4*: corresponding scattering spectra of different Au@Cu core–shell NPs in living HeLa cell (the colour bar in the scattering spectra indicates the wavelength of the maximum scattering intensity and reflects the resulting colour). **m** HRTEM image of a single Au@Cu core–shell nanoparticle. **n** Enlargement image of the Au@Cu core–shell nanoparticle. Reprinted with permission from Ref. [47]. Copyright (2011), WILEY-VCH Verlag GmbH & Co. KGaA, Weinheim

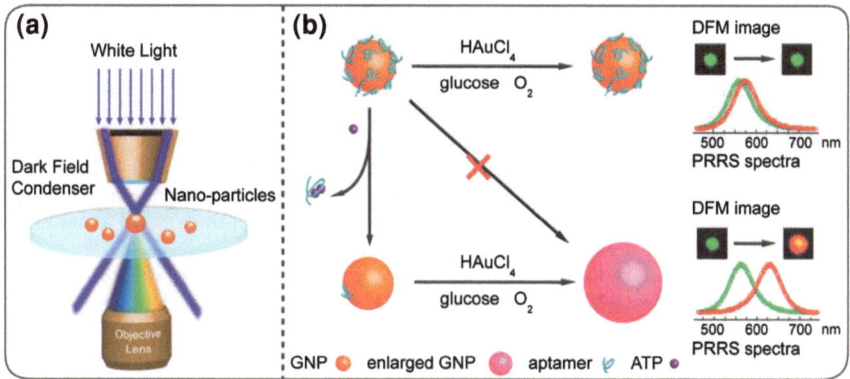

Fig. 4.19 a Experimental configuration of DFM and PRRS spectroscopy. **b** Schematic illustration of the aptasensor. The plasmonic signal is generated by the ATP-induced conformational change in surface adsorbed anti-ATP aptamer that recovers the self-catalytic activities of GNP. Reprinted with permission from Ref. [49]. Copyright (2012), Royal Society of Chemistry

plasmonic particles were treated with the drug taxol, which could confine the formation of NADH, their LSPR bands showed no obvious changes compared with Fig. 4.18l. This approach provides a new analytical tool to map biomolecules in cells and monitor metabolism processes in real time.

Fan [48] found that gold nanoparticles, in addition to exhibiting LSPR properties, also exhibit glucose oxidase (GOx)-like catalytic activity. GNP-catalysed glucose oxidation produces H_2O_2, which can reduce $HAuCl_4$ into gold atoms. Thus, in the in situ presence of glucose and $HAuCl_4$, gold nanoparticles increase in size. Further studies have indicated that the catalytic ability of gold nanoparticles is surface sensitive and that their catalytic activity is confined when molecules adsorb onto the particle surface. Researchers have used this phenomenon to develop single-nanoparticle biosensors that can be used to detect DNA hybridisation and conformational changes. As shown in Fig. 4.19, after ATP aptamers adsorbed onto gold nanoparticles, their catalytic ability was blocked and the particles exhibited no clear changes [49]. However, in the presence of ATP, the aptamer interacted with ATP and was removed from the particle surface, inducing particle growth. Using this method, the concentration of ATP can be detected on the basis of the scattering spectral shift of the GNPs. Furthermore, this biosensor can also be used to detect DNA hybridisation. Because single-stranded DNA is flexible and can adsorb onto the particle surfaces, the catalytic ability of the particles is confined. After the hybridisation with complementary DNA, the rigid double-stranded DNA was desorbed into solution, resulting in an increase in the particle size.

In conclusion, plasmonic nanoparticles with different morphologies and compositions exhibit variable optical, chemical, physical, and catalytic properties with multifunctions and benefits in wide applications. We can construct sensors by modulating their structures and components for ultrasensitive detection.

References

1. Link S, El-Sayed MA (2000) Shape and size dependence of radiative, non-radiative and photothermal properties of gold nanocrystals. Int Rev Phys Chem 19:409–453
2. Kelly KL, Coronado E, Zhao LL, Schatz GC (2003) The optical properties of metal nanoparticles: the influence of size, shape, and dielectric environment. J Phys Chem B 107:668–677
3. Jain PK, Lee KS, El-Sayed IH, El-Sayed MA (2006) Calculated absorption and scattering properties of gold nanoparticles of different size, shape, and composition: applications in biological imaging and biomedicine. J Phys Chem B 110:7238–7248
4. Gans R (1912) Über die form ultramikroskopischer goldteilchen. Ann Phys 37:881–900
5. Perez-Juste J, Pastoriza-Santos I, Liz-Marzan LM, Mulvaney P (2005) Gold nanorods: synthesis, characterization and applications. Coord Chem Rev 249:1870–1901
6. Hu M, Hillyard P, Hartland GV, Kosel T, Perez-Juste J, Mulvaney P (2004) Determination of the elastic constants of gold nanorods produced by seed mediated growth. Nano Lett 4:2493–2497
7. Gao J, Bender CM, Murphy CJ (2003) Dependence of the gold nanorod aspect ratio on the nature of the directing surfactant in aqueous solution. Langmuir 19:9065–9070
8. Huang X, El-Sayed IH, Qian W, El-Sayed MA (2006) Cancer cell imaging and photothermal therapy in the near-infrared region by using gold nanorods. J Am Chem Soc 128:2115–2120
9. Nelayah J, Kociak M, Stéphan O, de Abajo FJG, Tencé M, Henrard L et al (2007) Mapping surface plasmons on a single metallic nanoparticle. Nat Phys 3:348–353
10. Pedano ML, Li S, Schatz GC, Mirkin CA (2010) Periodic electric field enhancement along gold rods with nanogaps. Angew Chem 122:82–86
11. Heo CJ, Kim SH, Jang SG, Lee SY, Yang SM (2009) Gold "nanograils" with tunable dipolar multiple plasmon resonances. Adv Mater 21:1726–1731
12. Wang H, Brandl DW, Le F, Nordlander P, Halas NJ (2006) Nanorice: a hybrid plasmonic nanostructure. Nano Lett 6:827–832
13. McMahon JM, Wang Y, Sherry LJ, Van Duyne RP, Marks LD, Gray SK et al (2009) Correlating the structure, optical spectra, and electrodynamics of single silver nanocubes. J Phys Chem C 113:2731–2735
14. Becker J, Schubert O, Sönnichsen C (2007) Gold nanoparticle growth monitored in situ using a novel fast optical single-particle spectroscopy method. Nano Lett 7:1664–1669
15. Sherry LJ, Chang S-H, Schatz GC, Van Duyne RP, Wiley BJ, Xia Y (2005) Localized surface plasmon resonance spectroscopy of single silver nanocubes. Nano Lett 5:2034–2038
16. Mahmoud M, El-Sayed M (2011) Time dependence and signs of the shift of the surface plasmon resonance frequency in nanocages elucidate the nanocatalysis mechanism in hollow nanoparticles. Nano Lett 11:946–953

17. Hu M, Chen J, Marquez M, Xia Y, Hartland GV (2007) Correlated rayleigh scattering spectroscopy and scanning electron microscopy studies of Au-Ag bimetallic nanoboxes and nanocages. J Phys Chem C 111:12558–12565
18. Jin R, Cao Y, Mirkin CA, Kelly K, Schatz GC, Zheng J (2001) Photoinduced conversion of silver nanospheres to nanoprisms. Science 294:1901–1903
19. Härtling T, Alaverdyan Y, Wenzel MT, Kullock R, Käll M, Eng LM (2008) Photochemical tuning of plasmon resonances in single gold nanoparticles. J Phys Chem C 112:4920–4924
20. Haynes CL, Van Duyne RP (2001) Nanosphere lithography: a versatile nanofabrication tool for studies of size-dependent nanoparticle optics. J Phys Chem B 105:5599–5611
21. Polte J, Ahner TT, Delissen F, Sokolov S, Emmerling F, Thünemann AF et al (2010) Mechanism of gold nanoparticle formation in the classical citrate synthesis method derived from coupled in situ XANES and SAXS evaluation. J Am Chem Soc 132:1296–1301
22. Qin LX, Li Y, Li DW, Jing C, Chen BQ, Ma W et al (2012) Electrodeposition of single-metal nanoparticles on stable protein 1 membranes: application of plasmonic sensing by single nanoparticles. Angew Chem Int Ed 51:140–144
23. Jing C, Gu Z, Ying Y-L, Li D-W, Zhang L, Long Y-T (2012) Chrominance to dimension: a real-time method for measuring the size of single gold nanoparticles. Anal Chem 84:4284–4291
24. Song Y, Nallathamby PD, Huang T, Elsayed-Ali HE, Xu X-HN (2009) Correlation and characterization of three-dimensional morphologically dependent localized surface plasmon resonance spectra of single silver nanoparticles using dark-field optical microscopy and spectroscopy and atomic force microscopy. J Phys Chem C 114:74–81
25. Mock J, Barbic M, Smith D, Schultz D, Schultz S (2002) Shape effects in plasmon resonance of individual colloidal silver nanoparticles. J Chem Phys 116:6755–6760
26. Ringe E, Zhang J, Langille MR, Mirkin CA, Marks LD, Van Duyne RP (2012) Correlating the structure and localized surface plasmon resonance of single silver right bipyramids. Nanotechnology 23:444005–444011
27. Blaber MG, Henry A-I, Bingham JM, Schatz GC, Van Duyne RP (2011) LSPR imaging of silver triangular nanoprisms: correlating scattering with structure using electrodynamics for plasmon lifetime analysis. J Phys Chem C 116:393–403
28. Ringe E, Langille MR, Sohn K, Zhang J, Huang J, Mirkin CA et al (2012) Plasmon length: a universal parameter to describe size effects in gold nanoparticles. J Phys Chem Lett 3:1479–1483
29. Huang Y, Kim D-H (2011) Dark-field microscopy studies of polarization-dependent plasmonic resonance of single gold nanorods: rainbow nanoparticles. Nanoscale 3:3228–3232
30. Nehl CL, Liao H, Hafner JH (2006) Optical properties of star-shaped gold nanoparticles. Nano Lett 6:683–688
31. Anderson LJ, Payne CM, Zhen Y-R, Nordlander P, Hafner JH (2011) A tunable plasmon resonance in gold nanobelts. Nano Lett 11:5034–5037
32. Tang ML, Liu N, Dionne JA, Alivisatos AP (2011) Observations of shape-dependent hydrogen uptake trajectories from single nanocrystals. J Am Chem Soc 133:13220–13223
33. Nehl CL, Grady NK, Goodrich GP, Tam F, Halas NJ, Hafner JH (2004) Scattering spectra of single gold nanoshells. Nano Lett 4:2355–2359
34. Banholzer MJ, Harris N, Millstone JE, Schatz GC, Mirkin CA (2010) Abnormally large plasmonic shifts in silica-protected gold triangular nanoprisms. J Phys Chem C 114:7521–7526
35. Moon S, Kim Y, Oh Y, Lee H, Kim HC, Lee K et al (2012) Grating-based surface plasmon resonance detection of core-shell nanoparticle mediated DNA hybridization. Biosens Bioelectron 32:141–147
36. Srnová-Šloufová I, Vlčková B, Bastl Z, Hasslett TL (2004) Bimetallic (Ag) Au nanoparticles prepared by the seed growth method: two-dimensional assembling, characterization by energy dispersive X-ray analysis, X-ray photoelectron spectroscopy, and surface enhanced Raman spectroscopy, and proposed mechanism of growth. Langmuir 20:3407–3415
37. Wu Y, Jiang P, Jiang M, Wang T-W, Guo C-F, Xie S-S et al (2009) The shape evolution of gold seeds and gold@ silver core–shell nanostructures. Nanotechnology 20:305602–305612
38. Shore MS, Wang J, Johnston-Peck AC, Oldenburg AL, Tracy JB (2011) Synthesis of Au (Core)/Ag (Shell) nanoparticles and their conversion to AuAg alloy nanoparticles. Small 7:230–234

39. Deng J, Du J, Wang Y, Tu Y, Di J (2011) Synthesis of ultrathin silver shell on gold core for reducing substrate effect of LSPR sensor. Electrochem Commun 13:1517–1520
40. Xiong B, Zhou R, Hao J, Jia Y, He Y, Yeung ES (2013) Highly sensitive sulphide mapping in live cells by kinetic spectral analysis of single Au–Ag core-shell nanoparticles. Nat Commun 4:1708–1717
41. Park S-J, Duncan TV, Sanchez-Gaytan BL, Park S-J (2008) Bifunctional nanostructures composed of fluorescent core and metal shell subdomains with controllable geometry. J Phys Chem C 112:11205–11210
42. Raschke G, Brogl S, Susha A, Rogach A, Klar T, Feldmann J et al (2004) Gold nanoshells improve single nanoparticle molecular sensors. Nano Lett 4:1853–1857
43. Averitt R, Sarkar D, Halas N (1997) Plasmon resonance shifts of Au-coated Au$_2$S nanoshells: insight into multicomponent nanoparticle growth. Phys Rev Lett 78:4217–4220
44. Hien Pham TT, Cao C, Sim SJ (2008) Application of citrate-stabilized gold-coated ferric oxide composite nanoparticles for biological separations. J Magn Magn Mater 320:2049–2055
45. Peng S, Lei C, Ren Y, Cook RE, Sun Y (2011) Plasmonic/magnetic bifunctional nanoparticles. Angew Chem Int Ed 50:3158–3163
46. Xu Z, Hou Y, Sun S (2007) Magnetic core/shell Fe$_3$O$_4$/Au and Fe$_3$O$_4$/Au/Ag nanoparticles with tunable plasmonic properties. J Am Chem Soc 129:8698–8699
47. Zhang L, Li Y, Li DW, Jing C, Chen X, Lv M et al (2011) Single gold nanoparticles as real-time optical probes for the detection of NADH-dependent intracellular metabolic enzymatic pathways. Angew Chem 123:6921–6924
48. Zheng X, Liu Q, Jing C, Li Y, Li D, Luo W et al (2011) Catalytic gold nanoparticles for nanoplasmonic detection of DNA hybridization. Angew Chem 123:12200–12204
49. Liu Q, Jing C, Zheng X, Gu Z, Li D, Li D-W et al (2012) Nanoplasmonic detection of adenosine triphosphate by aptamer regulated self-catalytic growth of single gold nanoparticles. Chem Commun 48:9574–9576

Chapter 5
Interparticle Coupling-Enhanced Detection

Abstract When the distances between two or more plasmonic nanoparticles are very small, the plasmon resonance scattering spectra are greatly enhanced and distinct colour changes occur due to the coupling of the particles. Similar to fluorescence resonance energy transfer, plasmonic coupling is also distance dependent. Thus, researchers have fabricated colorimetric sensors by modulating the distance between nanoparticles, which have been used in a wide variety of applications, including DNA hybridisation, heavy-metal-ion detection, and protein binding. In this chapter, we primarily focus on the coupling of single particles, which enables the single-molecule detection through enhanced sensitivity.

Keywords Interparticle coupling • Chains of metal nanoparticles • Biosensors • Biomolecular detection • Cell imaging • Plasmonic nanopores

5.1 Fundamentals of Plasmonic Coupling

The coupling of plasmonics enhances the resonance intensity significantly and enables variable sensitive biosensors [1–11]. The factors that affect the coupling of nanoparticles include their particle size, coupling number, distance, direction, and shape [12–17]. Lee calculated the influence of particle size, number, composition, and distance using the generalised multiparticle Mie formalism [18]. The satellites in this work were made of gold with a relative permeability $\mu = 1$. Cores composed of either gold or glass were considered. The surrounding medium was assumed to be free space with a permittivity $\varepsilon = \mu = 1$. The permittivity of the glass core was 2.25. As shown in Fig. 5.1a, as the satellite number increased, the scattering peak wavelength exhibited a nearly linear redshift. For $r_{core} = 50$ nm, $r_{sat} = 10$ nm, and $d_{sat} = 2$ nm, the redshift was approximately 1 nm per satellite. Although these calculations may not accurately reflect absolute quantities, they reveal the trends in the peak wavelength as the number of satellites increases.

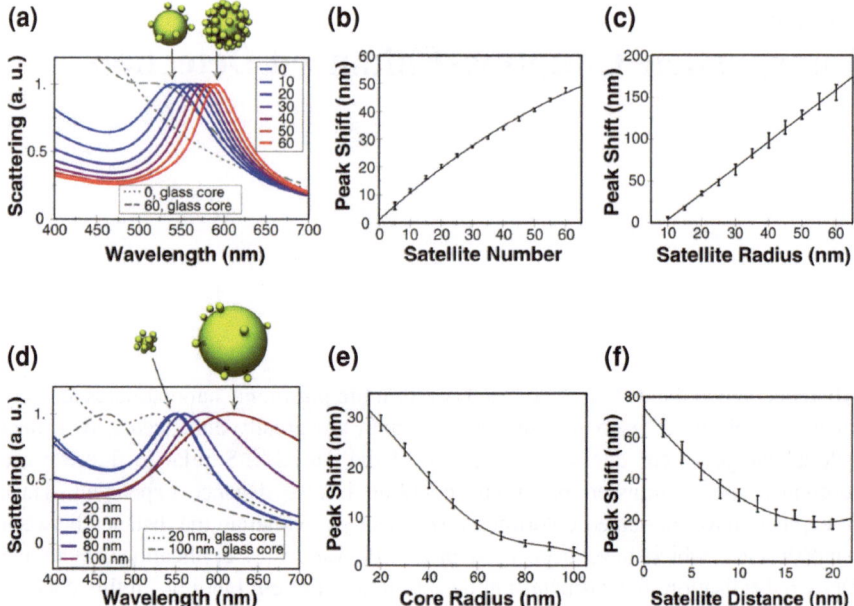

Fig. 5.1 Effect of varying satellite number, size and core size, satellite distance; normalised scattering cross section and peak shift $\Delta\lambda$ of core-satellite nanoassemblies for **a, b** increasing satellite number ($r_{core} = 50$ nm, $r_{sat} = 10$ nm, and $d_{sat} = 2$ nm); **c** increasing satellite radius ($r_{core} = 50$ nm, $n_{sat} = 5$, and $d_{sat} = 2$ nm); **d, e** increasing core radius ($r_{sat} = 10$ nm, $n_{sat} = 10$, and $d_{sat} = 2$ nm); and **f** increasing satellite distance ($r_{core} = 50$ nm, $r_{sat} = 30$ nm, and $n_{sat} = 5$); *solid line* is shown as a guide, and error bars represent standard deviation from ten randomly generated core-satellite assemblies. Reprinted with permission from Ref. [18]. Copyright (2009) AIP Publishing LLC

Meanwhile, the plasmon resonance bandwidth decreased with an increasing number of satellites because of local plasmonic coupling. The small nanoparticles reduced the damping effect, which decreased the coupling bandwidth. Glass cores with 60-nm gold nanoparticle satellites exhibited a broad scattering peak at approximately 550 nm, which was close to the scattering peak in the spectra of bare 50-nm gold nanoparticles. This result indicates that the plasmon resonance band shift and the narrowing of the bandwidth are primarily caused by the core-satellite interactions rather than by the satellite–satellite interactions. When the satellite radius was enlarged from 10 to 50 nm, as shown in Fig. 5.1c, an obvious redshift of 150 nm was obtained in the plasmon resonance band. The relationship between the plasmon resonance scattering peak shift and the size of the satellites was also approximately linear, with a 1-nm increase in the satellite radius leading to a 3-nm redshift. Notably, when the satellite radius was greater than 50 nm, the plasmon resonance band broadened substantially due to retardation effects because the size of the entire assembly was similar to the wavelength of incident light.

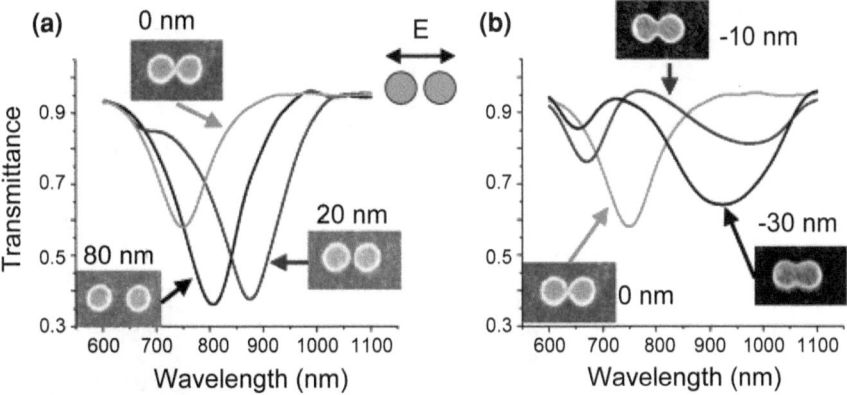

Fig. 5.2 Polarised transmission spectra in periodic arrays of pairwise interacting gold nanoparticles. The lattice constant is 800 nm in parallel direction to the pair axis, 400 nm perpendicular, and the dot height is 30 nm. **a** Pairs of particles approaching each other just reaching physical contact (labelled "0 nm" in the SEM image of a single-particle pair). **b** Increasing the particle overlap from point contact ("0 nm"; shown again for reference) and the widening of the interconnection "neck" into an eventually single ellipsoidal single (anisotropic) plasmonic particle. Reprinted with permission from Ref. [19]. Copyright (2004) American Chemical Society

In addition, the size of the core played an important role in the plasmon resonance band. As the core radius increased, the plasmon resonance band broadened and the peak shift decreased due to retardation effects. As shown in Fig. 5.1d, the plasmon resonance peak shift was approximately linear as the core radius increased (<50 nm) in the quasi-static regime. However, when the radius of the core was greater than 50 nm, the relationship between the peak shift and the radius increase was nonlinear due to intense damping effects, which caused the peak shift to decrease. In contrast, for the glass core without core-satellite coupling, the plasmon resonance peak was blueshifted. These results indicate that minimisation of the core radius is an effective method for increasing the peak shift. The distance d between the core and the satellite was subsequently investigated. In Fig. 5.1f, the red shift of plasmon band increased nonlinearly as the core-satellite decreased. The nonlinear relationship was caused by the rapidly decaying plasmonic near field, which decays as a function of d^{-3} in quasi-static approximations.

Nurmikko fabricated nanoparticle dimer arrays on ITO substrates to investigate dipole–dipole interactions in the conductively coupled regime, as shown in Fig. 5.2 [19]. Their results confirmed that the plasmon resonance of the nanoparticles exhibits a nonlinear relationship to increasing interparticle distance on a conductive substrate.

El-Sayed developed a "plasmon ruler equation" based on their investigation of lithographically fabricated gold nanodisc pairs with separation gaps of s = 212, 27, 17, and 2 nm using microabsorption spectroscopy and electrodynamic simulations [20]. The authors found that the plasmon resonance peak shift for polarisation along the interparticle axis shows nearly exponential decay with interparticle distance. The decay length is approximately 20 % of the particle size for different nanoparticle sizes, shapes, metal compositions, or medium dielectric constants. These results were explained

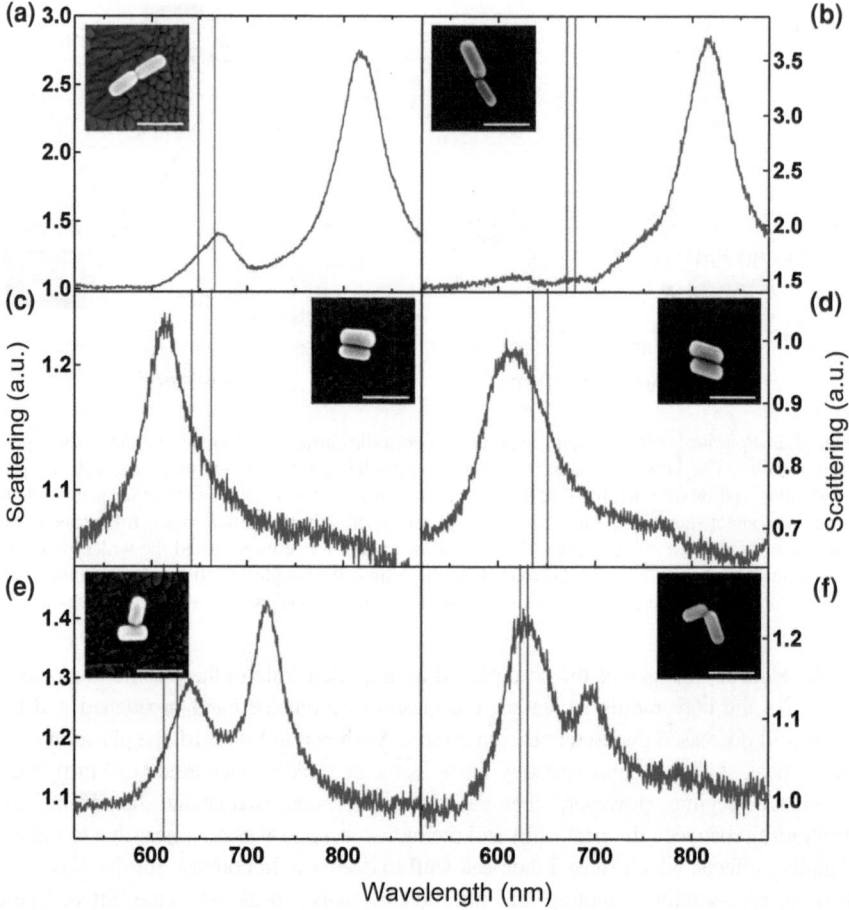

Fig. 5.3 Scattering spectrum for two rods aligned **a** and **b** end to end, **c** and **d** side to side, **e** in a T configuration and **f** in an L configuration, all on ITO and in air. *Insets* show the SEM images of the particles giving rise to each scattering spectrum. Scale bar = 100 nm. Reprinted with permission from Ref. [22]. Copyright (2009) American Chemical Society

by the dipolar coupling model because of two factors: the direct dependence of the single-particle polarisability on the cubic power of the particle dimension and the plasmonic near-field decay as the cubic power of the inverse distance. A "plasmon ruler equation" was therefore derived to estimate the interparticle coupling in a biological system, as shown in Eq. (5.1). In this equation, the refractive index for proteins is calculated to be approximately 1.6, $\Delta\lambda/\lambda_0$ is the fractional plasmon shift, s is the interparticle edge-to-edge separation, and D is the particle diameter.

$$\Delta\lambda/\lambda_0 \approx 0.18 \exp(-(s/d)/0.23) \tag{5.1}$$

Equation (5.1) is in good agreement with the experimental interparticle separation and can be applied for calibration of plasmon ruler design.

Fig. 5.4 Schematic diagrams of different manipulation techniques. **a** Nanoprobe manipulation. **b** Optical tweezers. **c** Plasmonic trapping. Reprinted with permission from Ref. [26]. Copyright (2013) Royal Society of Chemistry

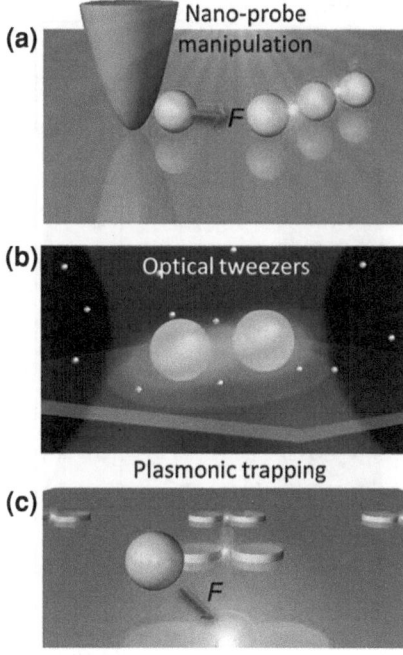

The geometries of nanorods also play an important role in plasmonic coupling [21, 22]. As shown in Fig. 5.3, the scattering spectra of two nanorods with different distances and angles were investigated. For the nanorods interacting end to end, as shown in Fig. 5.3a and b, the scattering peaks were redshifted by more than 130 nm relative to the peaks of individual rods attributed to the strong coupling of the longitudinal plasmon resonance modes. For the rods interacting side to side, the scattering peaks of the dimers exhibited a clear blueshift of 20–40 nm compared with the peaks of single rods because the low-intensity transverse plasmon modes interacted attractively. When the rods were arranged in L geometry, the coupled modes were arising from the interaction between the longitudinal plasmon modes rather than to longitudinal-transverse coupling. In contrast, the interaction coupling between the longitudinal and transverse modes affected the scattering cross section in the case of T geometry. DDA theory was used to calculate the energy distribution and plasmon scattering peaks for these interaction modes, and the results were in good agreement with the experimental data.

In addition to the coupling between single aggregated nanoparticles, asymmetric metal NP dimers, nanomanipulated metal NPs, and metal NPs supported on a substrate were investigated to improve our understanding of the underlying physics and to further extend coupling applications [23, 24]. The coupling of asymmetric metal NPs and nanomanipulated metal NPs provides good enhancement of scattered light and improved SERS detection, as shown in Fig. 5.4 [25–28].

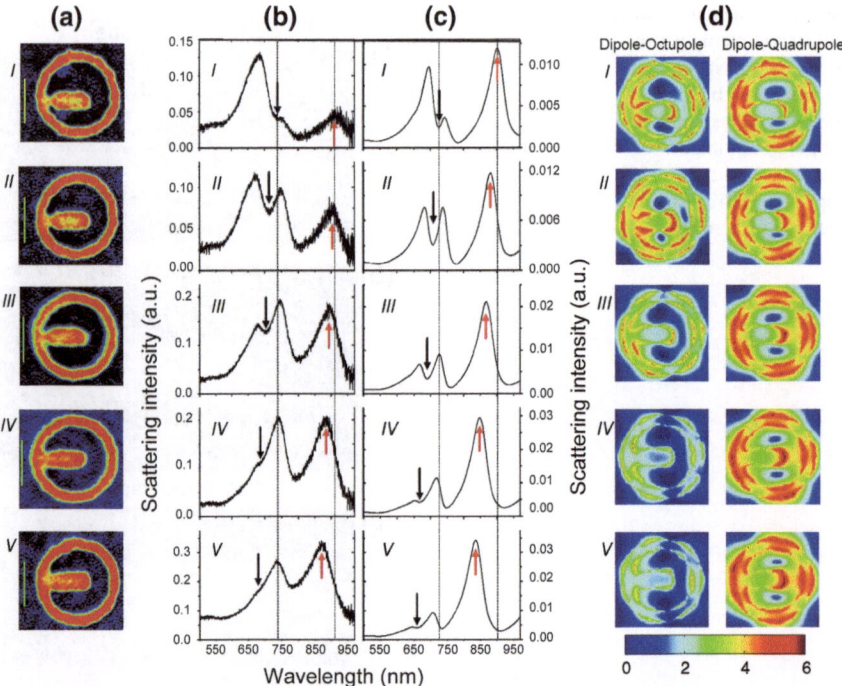

Fig. 5.5 Structural sensitivity of the quadrupolar and Fano resonances. **a** The SEM images are arranged in increasing order of the bridge size (scale bar = 100 nm). Generally, the width-to-radius aspect ratio ($\Delta R/R$) also increases in the same order due to fabrication properties. **b** Measured scattering spectra labelled the same as the corresponding images. **c** Calculated scattering cross section. The *red* and *black arrows* in **b** and **c** indicate the quadrupolar and Fano resonances, respectively. The wavelength shift of the quadrupolar and Fano resonances can be seen with reference to the *dotted vertical lines* in **b** and **c**, which show the position of the resonances for structure *I*. **d** The near-field intensity in natural logarithmic scale corresponding to the quadrupolar (*right*) and Fano (*left*) resonances, calculated 1 nm above the surface. Adapted with permission from Ref. [29]. Copyright (2011) American Chemical Society

The coupling of a NP with a metallic substrate is similar to the coupling between two adjacent NPs, with significant electromagnetic enhancement in the gap region. The hybridisation model in the nonretardation region can predict the shift in the plasmon resonance energy.

A theta-shaped ring-rod gold nanostructure whose resonance band could be modulated from 580 to 1,400 nm was fabricated, as shown in Fig. 5.5 [29]. The theta-shaped gold nanostructure enabled quadrupole plasmon and (octopolar) Fano resonances based on the coupling of the ring-rod. These resonances were dependent on the position and size of the rod, the width-to-radius aspect ratio ($\Delta R/R$), and the conductive bridge size. When the nanorod was in close proximity to the

Fig. 5.6 Low (*left*)- and high (*right*)-magnification transmission electron microscope images of nanoparticle dimers. **a** Symmetric dimers composed of 40-nm silver nanoparticles. **b** Asymmetric dimers composed of a 20- and a 40-nm silver nanoparticle. **c** Asymmetric dimers composed of a 30-nm silver nanoparticle and 40-nm gold nanoparticle. Reprinted with permission from Ref. [30]. Copyright (2010) American Chemical Society

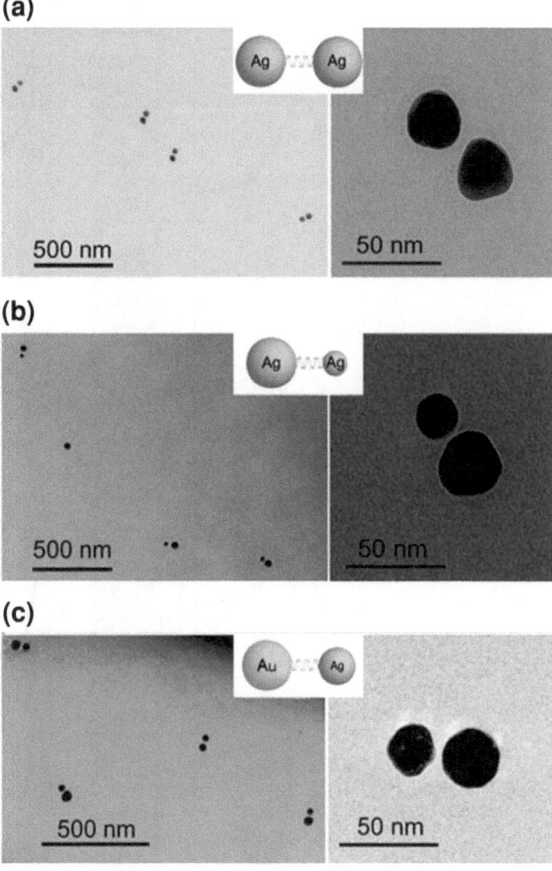

ring wall, a weak quadrupolar resonance induced by capacitive coupling was observed. As shown in Fig. 5.5, as the bridge size increased, the experimental and calculated spectra exhibited a blueshift in both the quadrupole plasmon and Fano resonance. In addition, an increase in $\Delta R/R$ caused the spectra to blueshift. When the nanorod was merged into a nanoring, the Fano resonance disappeared and the quadrupolar resonance became more pronounced. This ring-rod structure provided four resonances in response to the rod dipolar mode and the dipolar, quadrupolar, and octopolar modes of the ring.

Alivisatos studied the plasmonic coupling between asymmetric nanoparticle dimers (Fig. 5.6) [30], hexamers, and heptamers by gradually decreasing the interparticle gap separation (Fig. 5.7) [31]. They confirmed that both the gap distance and the plasmonic composition play important roles in near-field coupling.

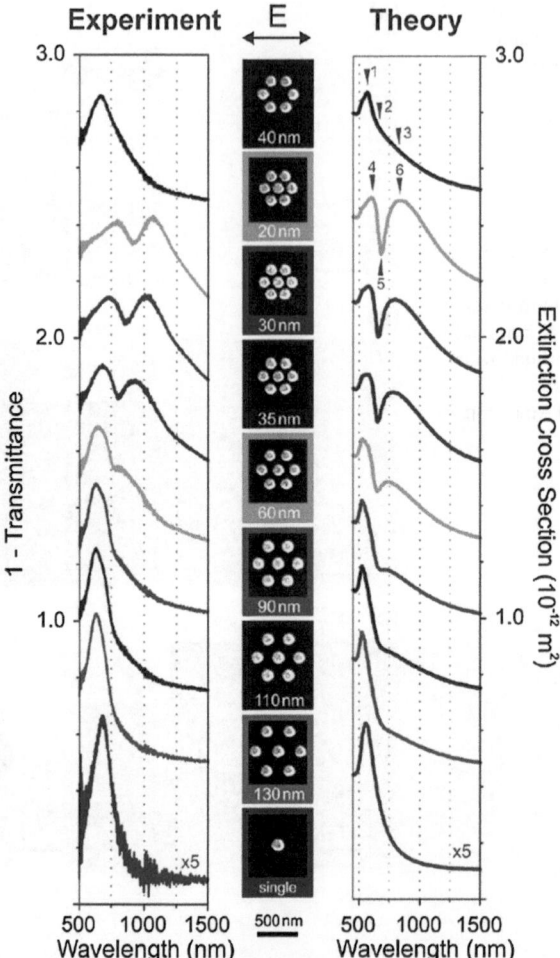

Fig. 5.7 Extinction spectra of a gold monomer, a gold hexamer, and gold heptamers with different interparticle gap separations. Spectra are shifted upward for clarity. *Left column*—the experimental extinction spectra (1-transmittance). *Middle column*—SEM images of the corresponding samples with indicated interparticle gap distances. The scale bar dimension is 500 nm. *Right column*—simulated extinction cross section spectra using the multiple multipole method. The gold structures are embedded in air. The difference between the experimental and simulated spectra is due to the presence of the glass substrate in the experiment, and it is also partially due to the assumption of a nanosphere shape for the trapezoidal nanoparticles in the simulation. In the gold monomer and hexamer, dipolar plasmon resonances are observed. The transition from isolated to collective modes is clearly visible in the different heptamers when decreasing the interparticle gap distance. Specifically, a pronounced Fano resonance is formed as characterised by the distinct resonance dip when the interparticle gap distance is below 60 nm. The presence or absence of the central nanoparticle can switch on or off the formation of the Fano resonance. Reprinted with permission from Ref. [31]. Copyright (2010) American Chemical Society

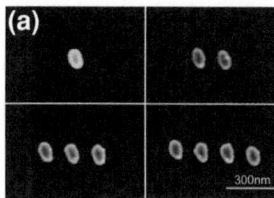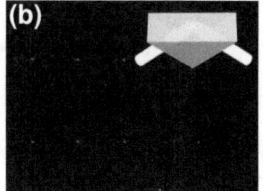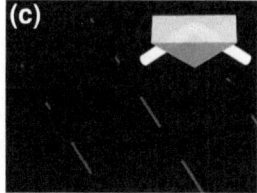

Fig. 5.8 **a** Typical SEM pictures of 1, 2, 3, and 4 particles in a line fabricated by standard e-beam lithography and lift-off process on a quartz substrate. The particles are elliptical with a 74 nm diameter in the short axis, a 102 nm diameter in the long axis, 30 nm thickness, and 153 nm centre–centre spacing. **b, c** Optical microscopic pictures taken with a digital camera for chains parallel **b** and at a 45° angle **c** with respect to the incident plane. The *inset* shows the geometrical configurations. Reprinted with permission from Ref. [36]. Copyright (2004) American Chemical Society

5.2 Coupling in Chains of Metal Nanoparticles

One-dimensional chains of metal nanoparticles have unique optical properties and can serve as "plasmon waveguides" on the basis of their near-field interparticle coupling [32–35]. Zhang fabricated finite one-dimensional chains of Au nanoparticles by electron beam lithography on quartz substrates and investigated their plasmonic scattering spectra experimentally and theoretically [36]. SEM images of the fabricated chains with 1–4 gold nanoparticles are shown in Fig. 5.8. The nanoparticles are measured 74 nm along their short axis and 102 nm along their long axis and are 30 nm thick. The centre-to-centre particle spacing is 153 nm, and the long axis is oriented at 74° with respect to the chain. The chains are separated by a 20 nm distance to avoid scattered light interference from other chains. To obtain the scattering images and spectra of the chains, a collimated light beam undergoing total internal reflection was used to produce evanescent light waves. A 150 W Xe white light source was delivered through a multimode optical fibre on a right-angled prism at a 45° angle. The scattering images of two types of chains with different orientations are shown in Fig. 5.9. Scattered light in the chains parallel to the incident plane of light was only observed at the two opposite ends of the chain, as shown in Fig. 5.8. When the chain was rotated to a 45° angle relative to the incident light, all the particles in the chain were observed as a red solid line under a dark-field microscope. These results were attributed to the interference effect caused by the periodicity of the particles in the chains. The light scattered from the chains was redshifted relative to that from single gold nanoparticles. The peak shift in the scattered light was dependent on the number of particles in the chain. For two particles, the plasmon resonance peak was redshifted approximately 30 nm because of interparticle coupling. For three particles, the scattered light peak only redshifted 14 nm compared with the peak of a single nanoparticle. For four particles, the peak was redshifted 25 nm, as shown in Fig. 5.9b. From both experiments and calculations, the authors concluded that the plasmon resonance peak shift was also dependent on the individual particle

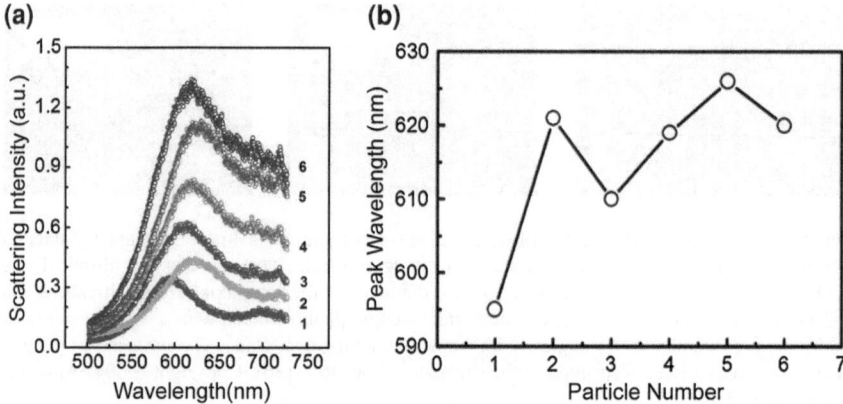

Fig. 5.9 **a** Measured scattering spectra for finite 1D chains of Au nanoparticles as exampled in Fig. 5.8a. **b** Peak resonance wavelength versus particle numbers in the chains. Reprinted with permission from Ref. [36]. Copyright (2004) American Chemical Society

size. For small particles, the red shift in the peak wavelength was nearly linear as the particle number increased. For large particles, the peak wavelength was not a monotonic function of particle number. In addition, the gap size between the nanoparticles in a chain substantially affected the plasmon resonance band. For a gap less than 50 nm, the peak shift for different particle numbers occurred in the order $3 > 2 > 5 > 4$, and for a gap between 50 and 60 nm, the sequence was $2 > 5 > 3 > 4$. This complex plasmonic behaviour could be attributed to the collaborative effect of both phase retardation and multiple plasmon resonance coupling.

A facile method for the preparation of sphere-to-string nanoparticle chains with spherical micellar shells, as shown in Fig. 5.10, was also reported [37]. In addition, the effects of particle morphology, composition, and spacing on the directional plasmon coupling were investigated by solution-phase self-assembly. Dark-field microscopy and polarised scattering (dark-field) microspectroscopy were used to characterise the directional near-field plasmonic coupling of a single chain. The scattering spectra of the chain (corresponding to the SEM image in Fig. 5.10) exhibited obvious differences in scattering intensity and peak wavelength under different polarisations of incident light, indicating directional plasmonic coupling.

5.3 Interparticle Coupling-Enhanced Biosensors

Because of the strong enhancement of plasmon resonance that results from interparticle coupling [38], Alivisatos used plasmonic nanoparticles to construct a type of molecular ruler to measure the length of DNA, as shown in Fig. 5.11 [39]. Two silver and gold nanoparticles were conjugated via a single strand of DNA modified with biotin and thiol. After the conjugation of DNA, the scattering spectra were redshifted approximately 50 nm for gold and 150 nm for silver compared with the individual

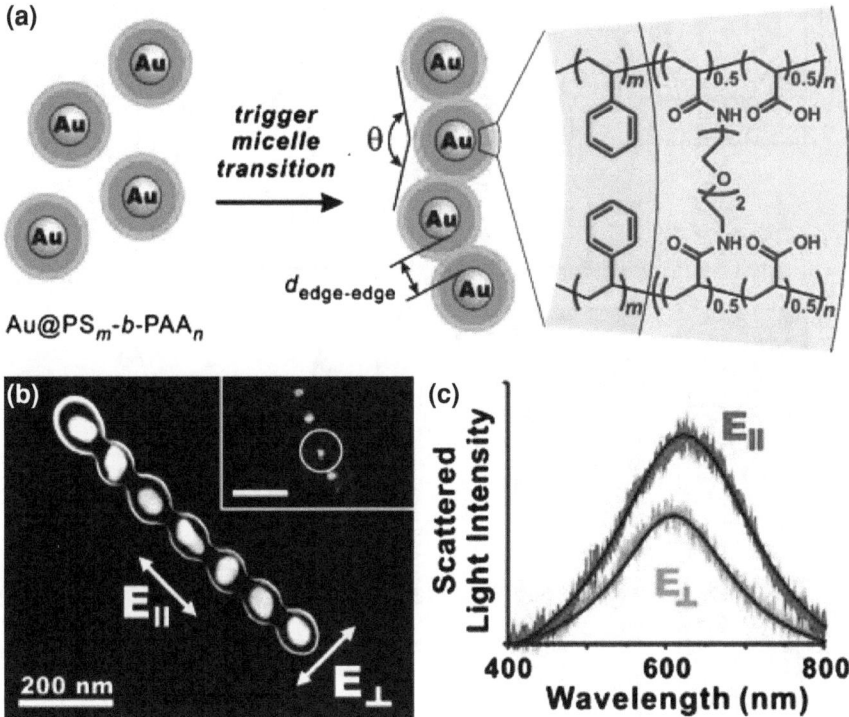

Fig. 5.10 a Synthesis process of nanochain. **b** SEM and dark-field optical microscopy (*inset*, scale bar represents 2 μm) images of a straight nanoparticle chain. The sample position in the optical image is rotated 25° clockwise relative to the SEM image. **c** Scattering spectra of the chain, collected under oblique illumination with light polarised parallel ($E_\parallel$) or perpendicular ($E_\perp$) to the long axis. Spectra were collected from the aperture-limited, *circled area* in the inset of **b**; λ_{max} ($E_\parallel$) 621 nm, λ_{max} ($E_\perp$) 607 nm. The *solid lines* are Lorentz fits of the spectral data. Adapted with permission from Ref. [37]. Copyright (2005) American Chemical Society

nanoparticles because of coupling effects. After the addition of complementary DNA, the hybridisation process was monitored in real time. Significant blue shifts in the scattering spectra of both the silver and gold nanoparticles were observed; these shifts were attributed to the greater stiffness of the dsDNA compared with that of the ssDNA, which led to the separation of the plasmonic nanoparticles. The peak shift of the dimers was in good agreement with the predicted length changes of the DNA linker. These plasmonic rulers enabled the monitoring of particle separation up to 70 nm for over 1 h.

Subsequently, the plasmonic rulers were used to monitor DNA cleavage catalysed by the EcoRV restriction enzyme [40]. EcoRV is a type II restriction endonuclease and transiently bends the DNA substrate. The bend angle is known to be 52° from crystal structures. As shown in Fig. 5.12, two gold nanoparticles were linked by DNA and attached to a glass slide. The DNA cleavage process catalysed by EcoRV was monitored through analysis of the intensity and wavelength of the light scattered by the dimer, and the bending of the DNA substrate due to interaction with the enzyme was confirmed.

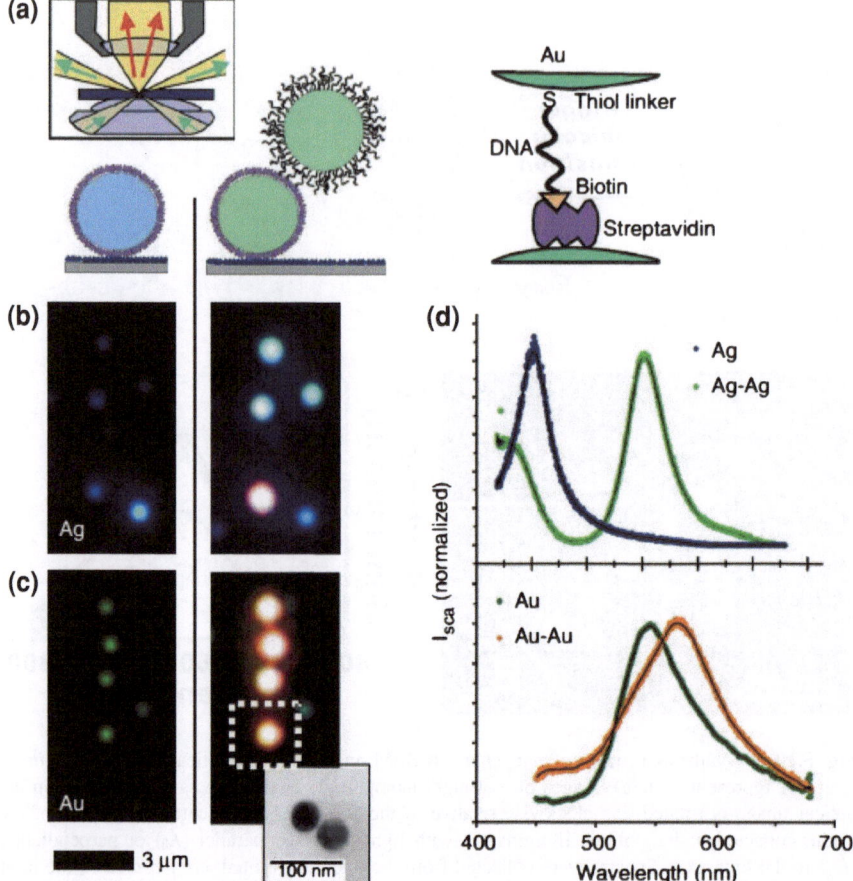

Fig. 5.11 Colour effect on directed assembly of DNA-functionalised gold and silver nanoparticles.
a First, nanoparticles functionalised with streptavidin are attached to the glass surface coated with
BSA-biotin (*left*). Then, a second particle is attached to the first particle (*centre*), again via biotin–streptavidin binding (*right*). The biotin on the second particle is covalently linked to the 3′ end
of a 33-bp-long ssDNA strand bound to the particle via a thiol group at the 5′ end. *Inset* principle
of transmission dark-field microscopy. **b** Single silver particles appear *blue* (*left*) and particle pairs
blue–green (*right*). The *orange dot* in the *bottom* comes from an aggregate of more than two particles. **c** Single gold particles appear *green* (*left*), and gold particle pairs appear *orange* (*right*). *Inset*
representative transmission electron microscopy image of a particle pair to show that each coloured
dot comes from light scattered from two closely lying particles, which cannot be separated optically. **d** Representative scattering spectra of single particles and particle pairs for silver (*top*) and
gold (*bottom*). Silver particles show a larger spectral shift (102 nm) than gold particles (23 nm),
stronger light scattering, and a smaller plasmon line width. Gold, however, is chemically more stable and is more easily conjugated to biomolecules via −SH, −NH$_2$, or −CN functional groups.
Reprinted with permission from Ref. [39]. Copyright (2005) Nature Publishing Group.

Utilising this phenomenon, researchers constructed a DNA sensor by combining
two gold nanoparticles through the hybridisation of DNA strands. As measured by

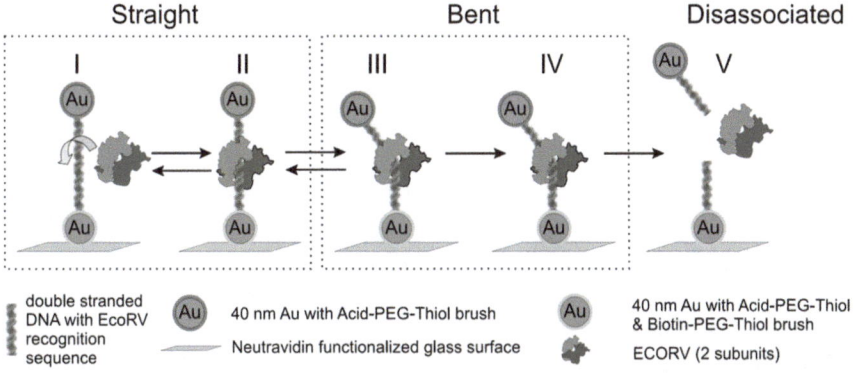

Fig. 5.12 Highly parallel single EcoRV restriction enzyme digestion assay. The plasmon rulers are immobilised with one particle to a glass surface through biotin–NeutrAvidin chemistry. The homodimeric EcoRV enzyme binds nonspecifically to DNA bound between the particles (*I*), translocates and binds to the target site (*II*), bends the DNA at the target site by ca. 50° (*III*), cuts the DNA in a blunt-ended fashion by phosphoryl transfer (54) (*IV*), and subsequently releases the products (*V*). Reprinted with permission from Ref. [40]. Copyright (2006) the National Academy of Sciences (USA)

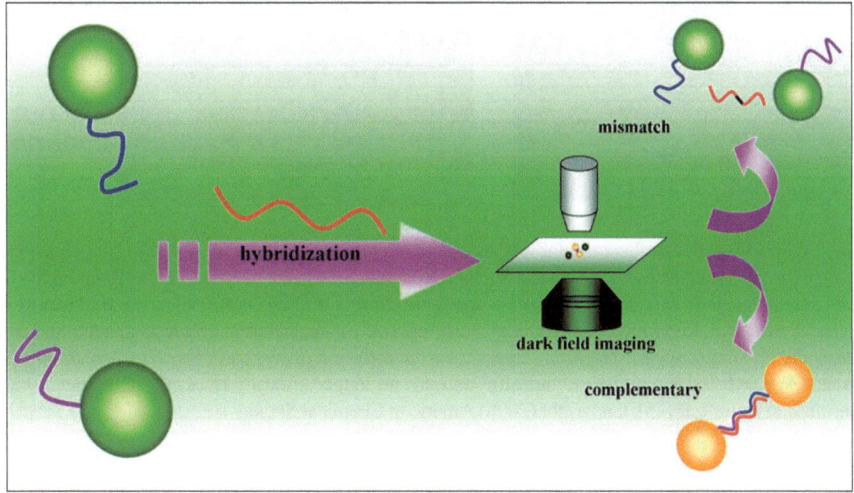

Fig. 5.13 Schematic diagram of the single-molecule sandwich assay with plasmonic resonant NPs. In the presence of target DNA, the monocolour NPs that are modified with cDNA probes would be pulled together upon target hybridisation. Plasmonic coupling between them causes the scattering spectrum of the dimer or oligomer to redshift from that of unbound NPs. Reprinted with permission from Ref. [42]. Copyright (2010) American Chemical Society

dark-field microscopy, colour changes in the light scattered by the gold nanoparticles indicated the DNA binding events and enabled the ultrasensitive detection of DNA, with a detection limit of 10^{-14} M [41]. To improve the selectivity and stability of the sensor, Au NP and Au/Ag/Au NPs were prepared to detect the binding of a single

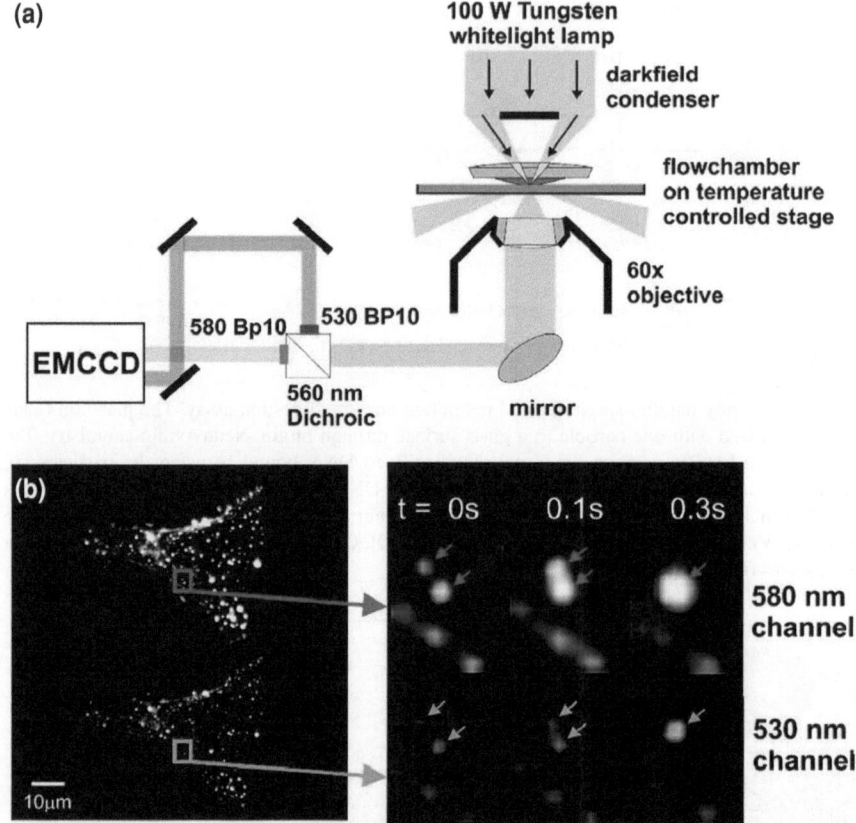

Fig. 5.14 a Experimental set-up. Individual gold nanoparticles are tracked in an inverted dark-field microscope. The collected light is chromatically separated, bandpass-filtered (580 BP10 and 530 BP 10), and captured on two translated areas of the same camera (EMCCD). **b** Image of a gold nanoparticle-labelled HeLa cell recorded simultaneously on two monochromatic colour channels: 580 nm (*top*) and 530 nm (*bottom*). The time series shows the diffusion of two particles on the plasma membrane. At $t = 0.3$ s, the particles colocalise and are no longer optically resolvable. Reprinted with permission from Ref. [45]. Copyright (2008) American Chemical Society

target DNA molecule between two NP probes without separation from the unbound NPs, as shown in Fig. 5.13 [42]. In addition, to enhance the sensitivity of the coupling effect between nanoparticles, a hairpin-loop DNA was used to increase the length extension after the DNA was hybridised [43, 44].

Reinhard utilised this near-field coupling property of plasmonic nanoparticles to measure subdiffraction limit distances in living cells, as shown in Fig. 5.14 [45]. Direct interactions between individual nanoparticle-labelled integrin surface receptors on living HeLa cells during colocalisation were investigated. Fibronectin was conjugated to integrins on the cell membranes by incubation for 10 min, and gold nanoparticles modified with antifibronectin were subsequently added. Real-time recording was initiated after the addition of the gold nanoparticles. When the

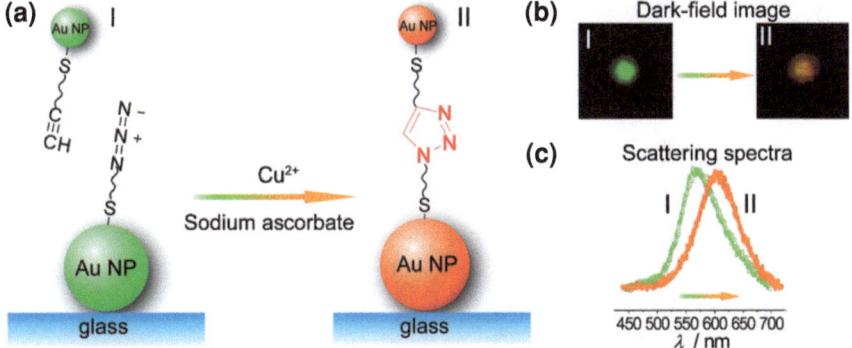

Fig. 5.15 a The detection of Cu^{2+} using click chemistry between two types of GNPs modified with terminal azide-functionalised and alkyne-functionalised thiols, respectively; Detailed experimental configuration. **b** A typical dark-field image of GNP modified on a microscopy slide before (*I*) and after (*II*) the addition of Cu^{2+} and sodium ascorbate. **c** Scattering spectra of single GNP before (*I*) and redshift after (II) the click reaction. Reprinted with permission from Ref. [46]. Copyright (2013) WILEY-VCH Verlag GmbH and Co. KGaA, Weinheim

particles approached each other within a distance of approximately one particle diameter, plasmonic coupling occurred. By this method, nanoparticles with very short separation distances (less than 100 nm) could be recognised much more sensitively than in conventional optical microscopy with a resolution of approximately 300–500 nm. This method provides a useful approach for monitoring the dynamic interactions of colocalised surface groups at the nanoscale.

Recently, Long and coworkers used the Cu^+-catalysed click reaction between alkyne and azide to link the alkyne-modified gold nanoparticles to an azide-modified 60-nm nanoparticle (Fig. 5.15) [46]. The maximum scattering wavelength of the single gold nanoparticle was significantly redshifted due to the plasmon resonance between the two kinds of gold nanoparticles. The formation of satellite nanoparticles allows for the real-time monitoring click reaction on single nanoparticles and sensitive detection of Cu^{2+}.

5.4 Plasmonic Nanopores

Recently, nanopores have attracted increasing attention as a potential DNA sequencing technique for detection at the single-molecule level [47–49]. The integration of nanopores and plasmonics provides a promising method for improving the detection efficiency of single molecules. Reuven Gordon and coworkers developed a new nanopore concept based on plasmonic coupling between nanoparticles and substrates, as shown in Fig. 5.16 [50]. They prepared two nanoholes on a 100-nm-thick Au film with a 2 nm Ti adhesion layer deposited by an e-beam onto a 1-inch square glass slide. When a nanoparticle enters the holes, the scattered light is

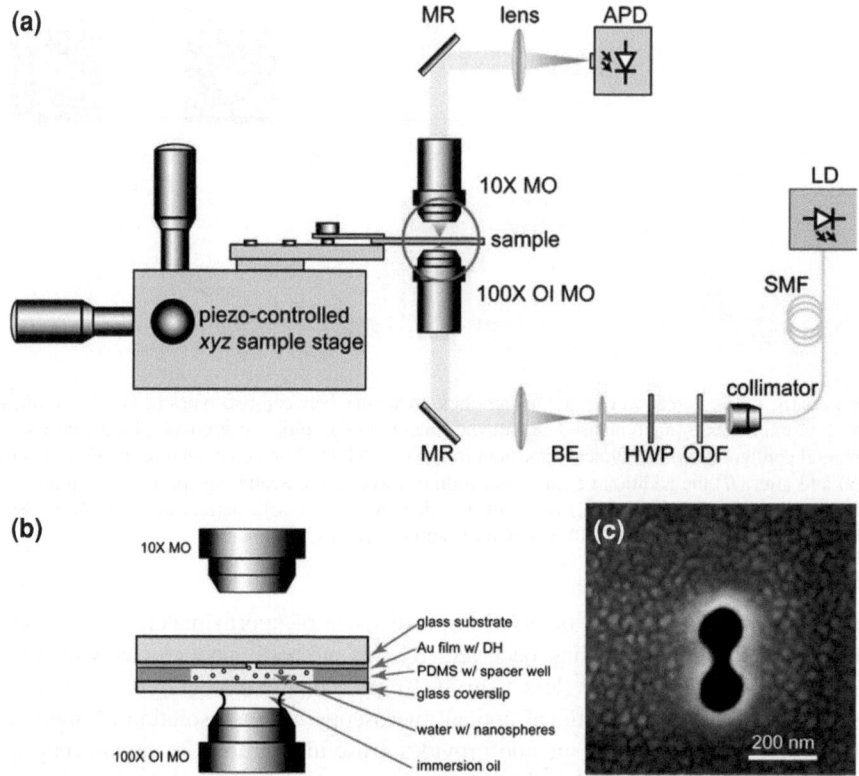

Fig. 5.16 **a** Schematic diagram of the nanoscale double-hole self-induced back-action optical trap. **b** An enlargement of the *circle* part in **a**, showing details of the composition of the sample in the microfluidic chamber, the set-up of the oil immersion microscope objective, and the condenser microscope objective. Abbreviations used: *LD* laser diode; *SMF* single-mode fibre; *ODF* optical density filter; *HWP* half-wave plate; *BE* beam expander; *MR* mirror; *MO* microscope objective; *OI MO* oil immersion microscope objective; *DH* double hole; *APD* avalanche photodetector. **c** SEM image of the double hole on Au film. Adapted with permission from Ref. [50]. Copyright (2011) American Chemical Society (Color figure online)

altered due to coupling. Thus, the passage of single nanoparticles can be recorded in real time. This method provides enhanced optical signal responses to the targets, whereas conventional nanopores provide weak electrical signals. Moreover, Sang-Hyun Oh developed a plasmonic nanopore using a gold nanohole array to detect the incorporation of a transmembrane protein, α-haemolysin (α-HL), as shown in Fig. 5.17 [51]. α-HL is a water-soluble peptide monomer (33.2 kD) secreted from the pathogenic bacteria *Staphylococcus aureus* that binds to the plasma membranes of numerous mammalian cell types. The periodic nanopore arrays were 200 nm in diameter and were based on Au/Si_3N_4 films with a 200-nm-thick Au film and a 20-nm silica layer. The nanopores array can act as a diffraction grating and convert incident light into surface plasmon resonance wavelengths to create intense peaks in the optical transmission spectra. This optical design provides multiplexing

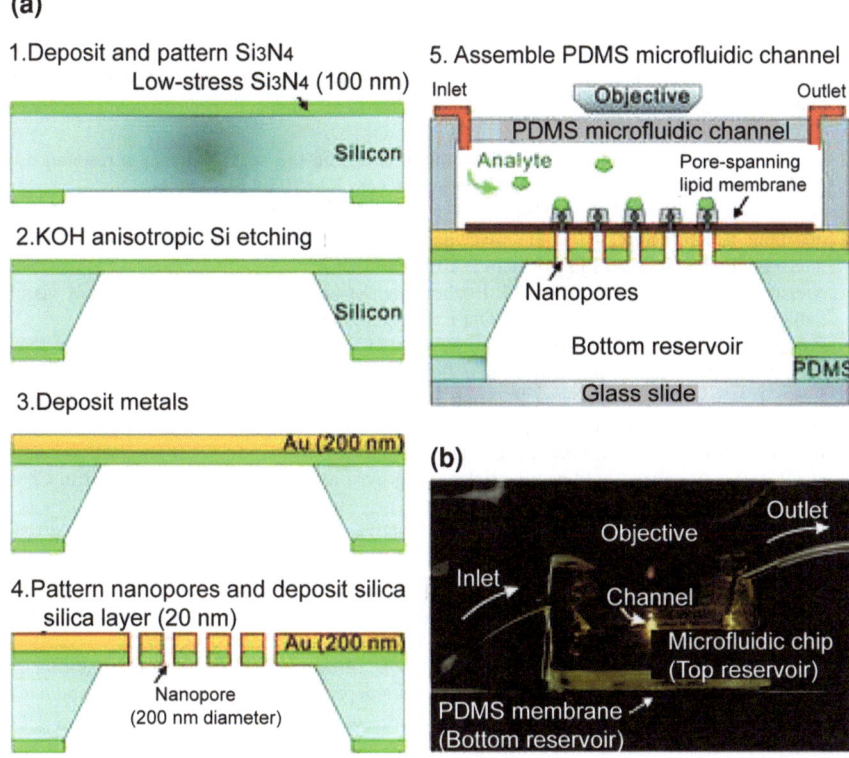

Fig. 5.17 **a** A process flow for making a free-standing nanopore array chip integrated with microfluidic channels. **b** A picture of the device on the microscope stage. A single-channel PDMS flow cell was attached on the top surface to inject analytes, and the reservoir on the back-side of the chip was scaled by a 300-μm-thick PDMS membrane to keep both sides of the pore-spanning lipid immersed in water. Reprinted with permission from Ref. [51]. Copyright (2010) Royal Society of Chemistry

capabilities and the potential for highly sensitive nanopore-based detection, which could be used to investigate membrane–protein interactions. A pore-spanning lipid membrane was then formed over the silica-coated Au surface of the nanopore array by vesicle rupture, mimicking natural cell membranes and causing a red shift in the resonance wavelength. After the addition of α-HL, kinetic measurements of α-HL binding on the membrane surface and the subsequent binding of anti-α-HL with α-HL were monitored in real time using the nanopore arrays. This method provides a natural platform for the real-time determination of cell membrane processes and expands the applications of plasmonic nanostructures.

In brief, the distance-dependent interparticle and particle–substrate couplings increase the plasmon resonance significantly, which offer an approach for the detection at single-molecule level and also can be used as a ruler to determine the molecule distance, length, and structure information.

References

1. Aizpurua J, Bryant GW, Richter LJ, García de Abajo FJ, Kelley BK, Mallouk T (2005) Optical properties of coupled metallic nanorods for field-enhanced spectroscopy. Phys Rev B 71:235420–235432
2. Ghosh SK, Pal T (2007) Interparticle coupling effect on the surface plasmon resonance of gold nanoparticles: from theory to applications. Chem Rev 107:4797–4862
3. Gunnarsson L, Rindzevicius T, Prikulis J, Kasemo B, Käll M, Zou S et al (2005) Confined plasmons in nanofabricated single silver particle pairs: experimental observations of strong interparticle interactions. J Phys Chem B 109:1079–1087
4. Nielsen MG, Pors A, Albrektsen O, Bozhevolnyi SI (2012) Efficient absorption of visible radiation by gap plasmon resonators. Opt Express 20:13311–13319
5. Frontiera RR, Gruenke NL, Van Duyne RP (2012) Fano-like resonances arising from long-lived molecule–plasmon interactions in colloidal nanoantennas. Nano Lett 12:5989–5994
6. Wang J, Wang L, Liu X, Liang Z, Song S, Li W et al (2007) A gold nanoparticle-based aptamer target binding readout for ATP assay. Adv Mater 19:3943–3946
7. Krpetić Z, Singh I, Su W, Guerrini L, Faulds K, Burley GA et al (2012) Directed assembly of DNA-functionalized gold nanoparticles using pyrrole-imidazole polyamides. J Am Chem Soc 134:8356–8359
8. Liu ZD, Li YF, Ling J, Huang CZ (2009) A localized surface plasmon resonance light-scattering assay of mercury (II) on the basis of Hg^{2+}-DNA complex induced aggregation of gold nanoparticles. Environ Sci Technol 43:5022–5027
9. Liu J, Lu Y (2005) Stimuli-responsive disassembly of nanoparticle aggregates for light-up colorimetric sensing. J Am Chem Soc 127:12677–12683
10. Elghanian R, Storhoff JJ, Mucic RC, Letsinger RL, Mirkin CA (1997) Selective colorimetric detection of polynucleotides based on the distance-dependent optical properties of gold nanoparticles. Science 277:1078–1081
11. Song Y, Xu X, MacRenaris KW, Zhang XQ, Mirkin CA, Meade TJ (2009) Multimodal gadolinium-enriched DNA-gold nanoparticle conjugates for cellular imaging. Angew Chem Int Ed 48:9143–9147
12. Chang W-S, Willingham BA, Slaughter LS, Khanal BP, Vigderman L, Zubarev ER et al (2011) Low absorption losses of strongly coupled surface plasmons in nanoparticle assemblies. Proc Natl Acad Sci 108:19879–19884
13. Zhang L, Chen H, Wang J, Li YF, Wang J, Sang Y et al (2010) Tetrakis (4-sulfonatophenyl) porphyrin-directed assembly of gold nanocrystals: tailoring the plasmon coupling through controllable gap distances. Small 6:2001–2009
14. Mastroianni AJ, Claridge SA, Alivisatos AP (2009) Pyramidal and chiral groupings of gold nanocrystals assembled using DNA scaffolds. J Am Chem Soc 131:8455–8459
15. Yang L, Wang H, Yan B, Reinhard BM (2010) Calibration of silver plasmon rulers in the 1–25 nm separation range: experimental indications of distinct plasmon coupling regimes. J Phys Chem C 114:4901–4908
16. Nordlander P, Oubre C, Prodan E, Li K, Stockman MI (2004) Plasmon hybridization in nanoparticle dimers. Nano Lett 4:899–903
17. Woo KC, Shao L, Chen H, Liang Y, Wang J, Lin H-Q (2011) Universal scaling and Fano resonance in the plasmon coupling between gold nanorods. ACS Nano 5:5976–5986
18. Ross BM, Waldeisen JR, Wang T, Lee LP (2009) Strategies for nanoplasmonic core-satellite biomolecular sensors: theory-based design. Appl Phys Lett 95:193112–193114
19. Atay T, Song J-H, Nurmikko AV (2004) Strongly interacting plasmon nanoparticle pairs: from dipole–dipole interaction to conductively coupled regime. Nano Lett 4:1627–1631
20. Jain PK, Huang W, El-Sayed MA (2007) On the universal scaling behavior of the distance decay of plasmon coupling in metal nanoparticle pairs: a plasmon ruler equation. Nano Lett 7:2080–2088
21. Shao L, Woo KC, Chen H, Jin Z, Wang J, Lin H-Q (2010) Angle- and energy-resolved plasmon coupling in gold nanorod dimers. ACS Nano 4:3053–3062

22. Funston AM, Novo C, Davis TJ, Mulvaney P (2009) Plasmon coupling of gold nanorods at short distances and in different geometries. Nano Lett 9:1651–1658
23. Wang X, Gogol P, Cambril E, Palpant B (2012) Near-and far-field effects on the plasmon coupling in gold nanoparticle arrays. J Phys Chem C 116:24741–24747
24. Yang L, Yan B, Reinhard BM (2008) Correlated optical spectroscopy and transmission electron microscopy of individual hollow nanoparticles and their dimers. J Phys Chem C 112:15989–15996
25. Jamshidi A, Pauzauskie PJ, Schuck PJ, Ohta AT, Chiou P-Y, Chou J et al (2008) Dynamic manipulation and separation of individual semiconducting and metallic nanowires. Nat Photonics 2:86–89
26. Tong L, Wei H, Zhang S, Li Z, Xu H (2013) Optical properties of single coupled plasmonic nanoparticles. Phys Chem Chem Phys 15:4100–4109
27. Mock JJ, Hill RT, Degiron A, Zauscher S, Chilkoti A, Smith DR (2008) Distance-dependent plasmon resonant coupling between a gold nanoparticle and gold film. Nano Lett 8:2245–2252
28. Vernon KC, Funston AM, Novo C, Gómez DE, Mulvaney P, Davis TJ (2010) Influence of particle-substrate interaction on localized plasmon resonances. Nano Lett 10:2080–2086
29. Habteyes TG, Dhuey S, Cabrini S, Schuck PJ, Leone SR (2011) Theta-shaped plasmonic nanostructures: bringing "dark" multipole plasmon resonances into action via conductive coupling. Nano Lett 11:1819–1825
30. Sheikholeslami S, Jun Y-W, Jain PK, Alivisatos AP (2010) Coupling of optical resonances in a compositionally asymmetric plasmonic nanoparticle dimer. Nano Lett 10:2655–2660
31. Hentschel M, Saliba M, Vogelgesang R, Giessen H, Alivisatos AP, Liu N (2010) Transition from isolated to collective modes in plasmonic oligomers. Nano Lett 10:2721–2726
32. Brongersma ML, Hartman JW, Atwater HA (2000) Electromagnetic energy transfer and switching in nanoparticle chain arrays below the diffraction limit. Phys Rev B 62:R16356–R16359
33. Solis DJ, Willingham B, Nauert SL, Slaughter LS, Olson J, Swanglap P et al (2012) Electromagnetic energy transport in nanoparticle chains via dark plasmon modes. Nano Lett 12:1349–1353
34. Grzelczak M, Mezzasalma SA, Ni W, Herasimenka Y, Feruglio L, Montini T et al (2012) Antibonding plasmon modes in colloidal gold nanorod clusters. Langmuir 28:8826–8833
35. Février M, Gogol P, Aassime A, Mégy R, Delacour Cc, Chelnokov A et al (2012) Giant coupling effect between metal nanoparticle chain and optical waveguide. Nano Lett 12:1032–1037
36. Wei Q-H, Su K-H, Durant S, Zhang X (2004) Plasmon resonance of finite one-dimensional Au nanoparticle chains. Nano Lett 4:1067–1071
37. Kang Y, Erickson KJ, Taton TA (2005) Plasmonic nanoparticle chains via a morphological, sphere-to-string transition. J Am Chem Soc 127:13800–13801
38. Wang H, Reinhard BM (2009) Monitoring simultaneous distance and orientation changes in discrete dimers of DNA linked gold nanoparticles. J Phys Chem C 113:11215–11222
39. Sönnichsen C, Reinhard BM, Liphardt J, Alivisatos AP (2005) A molecular ruler based on plasmon coupling of single gold and silver nanoparticles. Nat Biotechnol 23:741–745
40. Reinhard BM, Sheikholeslami S, Mastroianni A, Alivisatos AP, Liphardt J (2007) Use of plasmon coupling to reveal the dynamics of DNA bending and cleavage by single EcoRV restriction enzymes. Proc Natl Acad Sci 104:2667–2672
41. Yuan Z, Cheng J, Cheng X, He Y, Yeung ES (2012) Highly sensitive DNA hybridization detection with single nanoparticle flash-lamp darkfield microscopy. Analyst 137:2930–2932
42. Xiao L, Wei L, He Y, Yeung ES (2010) Single molecule biosensing using color coded plasmon resonant metal nanoparticles. Anal Chem 82:6308–6314
43. Sebba DS, Mock JJ, Smith DR, Labean TH, Lazarides AA (2008) Reconfigurable core-satellite nanoassemblies as molecularly-driven plasmonic switches. Nano Lett 8:1803–1808
44. Verdoold R, Gill R, Ungureanu F, Molenaar R, Kooyman RP (2011) Femtomolar DNA detection by parallel colorimetric darkfield microscopy of functionalized gold nanoparticles. Biosens Bioelectron 27:77–81

45. Rong G, Wang H, Skewis LR, Reinhard BM (2008) Resolving sub-diffraction limit encounters in nanoparticle tracking using live cell plasmon coupling microscopy. Nano Lett 8:3386–3393

46. Shi L, Jing C, Ma W, Li DW, Halls JE, Marken F, Long YT (2013) Plasmon resonance scattering spectroscopy at the single-nanoparticle level: real-time monitoring of a click reaction. Angew Chem Int Ed 52:6011–6014

47. Branton D, Deamer DW, Marziali A, Bayley H, Benner SA, Butler T et al (2008) The potential and challenges of nanopore sequencing. Nat Biotechnol 26:1146–1153

48. Ying YL, Li DW, Li Y, Lee JS, Long YT (2011) Enhanced translocation of poly(dt)$_{45}$ through an α-hemolysin nanopore by binding with antibody. Chem Commun 47:5690–5692

49. Ying YL, Wang HY, Sutherland TC, Long YT (2011) Monitoring of an ATP-binding aptamer and its conformational changes using an α-hemolysin nanopore

50. Pang Y, Gordon R (2011) Optical trapping of 12 nm dielectric spheres using double-nanoholes in a gold film. Nano Lett 11:3763–3767

51. Im H, Wittenberg NJ, Lesuffleur A, Lindquist NC, Oh S-H (2010) Membrane protein biosensing with plasmonic nanopore arrays and pore-spanning lipid membranes. Chem Sci 1:688–696

Chapter 6
Detection Based on Plasmon Resonance Energy Transfer

Abstract For nanoparticles with adsorbed chromophores, when the absorption bands of chromophores are overlapped with the resonance scattering bands of particles, "plasmon resonance energy transfer" (PRET) from metal nanoparticles to the surface-modified chromophores occurs. PRET enhances the sensitivity of absorption signals of chromophores with several orders of magnitudes. In this chapter, we discuss the discovery of PRET as well as its applications in ultrasensitive sensors.

Keywords Plasmon resonance energy transfer • Chromophores • Absorption spectroscopy • Cytochrome c • Cellular imaging • Heavy metal ions

Although optical absorption spectroscopy is a widely used analytical method in chemistry, it suffers from some shortcomings, such as low sensitivity and low spatial resolution, which seriously limit its applications in ultrasensitive biomolecular analysis and in vivo cellular imaging. The combination of optical absorption spectroscopy with single-particle Rayleigh scattering might provide enhanced sensitivity and improve the spatial resolution of absorption methods to the nanoscale, which provides a new path for in vivo imaging with chemical fingerprint information [1–6].

The phenomenon of "plasmon resonance energy transfer" (PRET) was first reported by Lee et al. [7] in 2007, who successfully achieved this integration of absorption and single-particle scattering. When nanoparticles are coupled to molecular chromophores, the plasmon resonance energy can be transferred from the nanoparticles to the molecules, which will quench the Rayleigh-scattering spectrum if the resonance band of the nanoparticle overlaps with the chromophore absorption band. The quenching position corresponds to the absorption peak of the molecular dye, which acts as the energy accepter. Before it was experimentally observed, this phenomenon had already been hypothesised to account for the surface-enhanced Raman scattering, fluorescence, and luminescence of single nanoparticles [8–10]. Lee and his group discovered quenching dips in the spectra of plasmon resonance

Y.-T. Long and C. Jing, *Localized Surface Plasmon Resonance Based Nanobiosensors*, SpringerBriefs in Molecular Science, DOI: 10.1007/978-3-642-54795-9_6,
© The Author(s) 2014

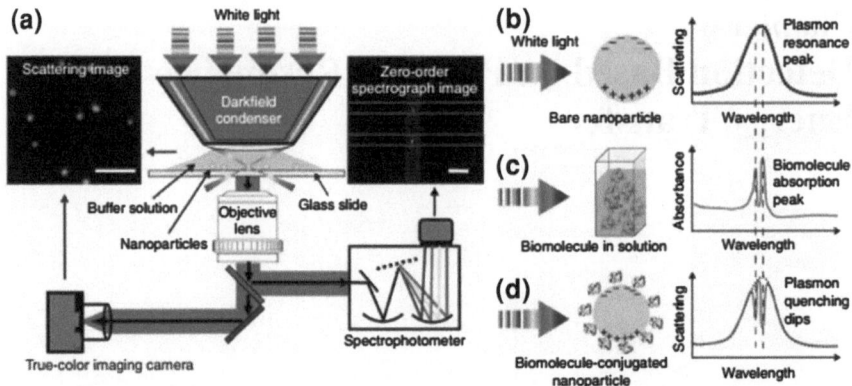

Fig. 6.1 Schematic diagrams of quantised plasmon quenching dips nanospectroscopy via PRET. **a** Experimental system configuration. **b** A typical Rayleigh-scattering spectrum of bare gold nanoparticles. **c** Typical absorption spectra of biomolecule bulk solution. **d** Typical quantised plasmon quenching dips in the Rayleigh-scattering spectrum of biomolecule-conjugated gold nanoparticles. Spectra were drawn based on representative data. Reprinted with permission from Ref. [7]. Copyright (2007) Nature Publishing Group

Rayleigh-scattered light from single nanoparticles which induced by the direct quantised PRET from the nanoparticle to the biomolecules, such as cytochrome *c* (Cyt. C), adsorbed onto its surface (Fig. 6.1).

Lee's work demonstrated that, when metal nanoparticles are conjugated to reduced or oxidised Cyt. C, distinct quenching is observed in the LSPR scattering spectrum between 520 and 550 nm, which overlaps the two absorption peaks from the redox states of Cyt. C in solution. However, when the LSPR spectrum did not overlap with the absorption band of Cyt. C, neither quenching dips nor any detectable signals for dielectric polystyrene nanoparticle–cytochrome *c* conjugates (Fig. 6.2i) and nanoparticle–peptide conjugates (Fig. 6.2g) were observed.

Although the mechanism is not yet fully understood, one possible explanation is that the energy is transferred via dipole–dipole interactions between the plasmon resonance dipole of the nanoparticles and the absorption dipole of the chromophore moiety, which is similar to the fluorescence resonance energy transfer (FRET) mechanism. Several previous studies on the surface plasmon-mediated FRET process [11], the surface plasmon resonance shift due to adsorbed redox molecules [12], and the bulk optical extinction spectroscopy of nanoplasmonic particles with conjugated resonant molecules [13] could provide evidence for such explanation.

While the gold nanoparticles are already recognised as an ideal material for cellular imaging because of their outstanding water solubility, biocompatibility, and low photobleaching and photoblinking, the PRET technique provides a powerful new tool for label-free biomolecular analysis, especially for in vivo cellular and molecular imaging. In a cell, Cyt. C is released from the mitochondria to the cytoplasm to increase the permeability of the outer membrane of the mitochondria in response to pro-apoptotic stimuli. The real-time molecular imaging of ethanol

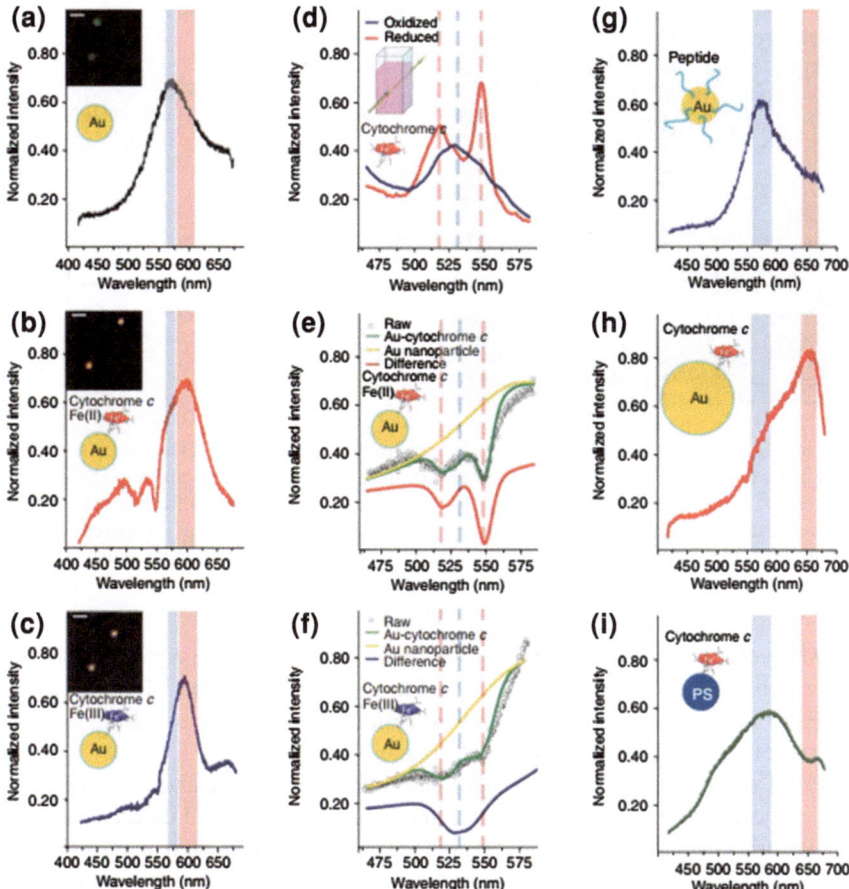

Fig. 6.2 Experimental results of PRET from single gold nanoparticle to conjugated cytochrome *c* molecules. **a–c** The Rayleigh-scattering spectra (obtained using 1-s integration time) of three 30-nm gold nanoparticles coated with cysteamine only (**a**), cysteamine and reduced cytochrome *c* (**b**), and cysteamine and oxidised cytochrome *c* (**c**). *Insets*, true-colour scattering images of individual nano-particles. Scale bars, 2 mm. **d** The bulk visible absorption spectra of oxidised (*blue solid line*) and reduced (*red solid line*) 8-mM cytochrome *c* using conventional UV–vis absorption spectroscopy. *Dashed lines* were drawn to facilitate identification of peak wavelengths. **e** *Fitting curve* for the spectrum in (**b**). *Green solid line*, fitting curve of the raw data. *Yellow solid line*, Lorentzian *scattering curve* of bare gold nanoparticle. *Red solid line*, processed absorption spectra for reduced con-jugated cytochrome *c* (*green minus yellow curve*). **f** The *fitting curve* for the spectrum in c. *Blue solid line*, processed absorption spectra for oxidised conjugated cytochrome *c* (*green minus yel-low curve*). **g–i** PRET spectra of gold nanoparticle coated with Cys-(Gly-Hyp-Pro)$_6$ peptide (**g**), cytochrome *c* on large gold nanoparticle cluster (**h**) and cytochrome *c* on a 40-nm polystyrene nan-oparticle (**i**). Reprinted with permission from Ref. [7]. Copyright (2007) Nature Publishing Group

induced Cyt. C production in living cells has been achieved through the application of PRET [14]. In this work, no quenching was observed prior to exposure of the cell to ethanol. After the stimuli were applied, a significant quenching dip at 550 nm

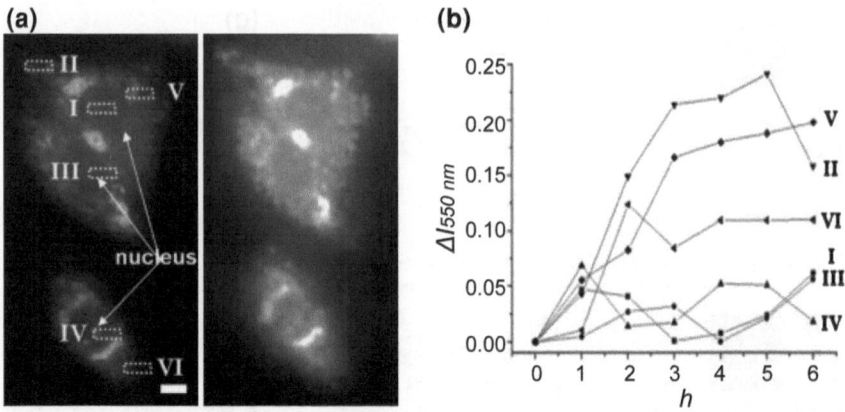

Fig. 6.3 a *Left* Representative *dark-field image* (from CCD camera) of HepG2 cells labelled with carboxylic acid-terminated probes prior to exposure with 100 mM ethanol. The *bar* corresponds to 10 μm. *Right* Corresponding *B/W dark-field scattering image* from spectrometer. **b** Time-course of the differential quenching dip changes at different positions compared to values at 0 h. Reprinted with permission from Ref. [14]. Copyright (2009) American Chemical Society

could be observed. The wavelength-specific quenching dips at positions I, III, and IV, which correspond to the nucleus, and at positions II, V, and VI, which correspond to the plasma, were monitored over time; the results revealed the dynamic process of Cyt. C production in the cell (Fig. 6.3).

PRET could also be used to design sensors for the selective and sensitive detection of metal ions based on a metal–ligand complex, which would induce absorption band overlap with the scattering peak of the plasmonic nanostructure (Fig. 6.4).

Lee and coworkers [15] reported a PRET-based ion-sensing technique to detect concentrations of Cu^{2+} as low as 1 nM and applied their technique to detect the changes in intracellular Cu^{2+} in living HeLa cells. The amine complex of Cu^{2+} has an optical absorption peak in the visible range at approximately 550 nm, which coincides with the scattering peak of 50-nm gold nanoparticles. The surfaces of gold nanoparticles were modified with ethylenediamine moieties for the selective recognition of aqueous Cu^{2+} ions. The authors evaluated the selectivity of the Cu^{2+} sensor by testing its response to other metal ions, no obvious intensity variations in the scattering peak were observed for any metal ions except Cu^{2+}.

Peak quenching depends on both the shape of the peak in the scattering spectrum and the width of the absorption band. As observed previously, when the fwhm of the absorption band was significantly shorter than that of the scattering peak of the plasmonic probe, as in the case of Cyt. C, a "spectral quenching dip" was observed as the result of PRET. In this work, the Cu^{2+} complex exhibits a broad absorption band with a fwhm greater than 90 nm, which results in a "spectral decrease" rather than a "quenching dip" (Fig. 6.4).

The sensitivity of absorption spectroscopy was enhanced by PRET by several orders of magnitude, thereby allowing the detection of just several hundred

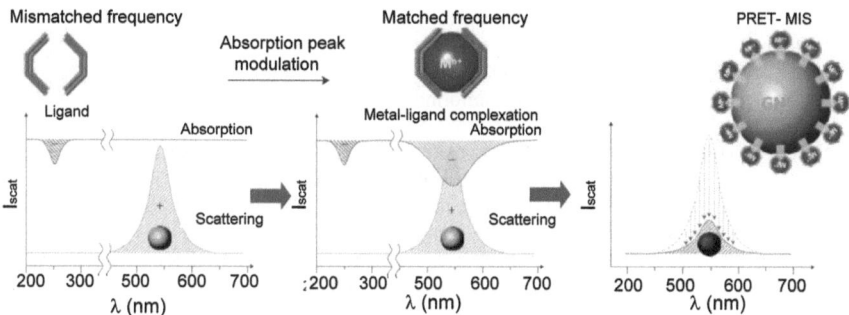

Fig. 6.4 There is no spectral overlap between ligands without the metal ion and the GNP (*left*). When the electronic absorption frequency of the metal–ligand complex matches with the Rayleigh-scattering frequency, the selective energy transfer is induced by this spectral overlap (*middle*) and the distinguishable resonant quenching on the resonant Rayleigh-scattering spectrum is observed (*right*). Reprinted with permission from Ref. [15]. Copyright (2009) Nature Publishing Group

molecules. PRET biosensors enable sensitivity as much as 100–1,000 times [15] greater than that of traditional organic report-based methods as well as high spatial resolution due to the nanoscale size of the particle-based sensor. Benefitting from this sensitivity and special resolution, PRET could be widely used for biosensor and Rayleigh-scattering imaging.

References

1. Bruchez M, Moronne M, Gin P, Weiss S, Alivisatos AP (1998) Semiconductor nanocrystals as fluorescent biological labels. Science 281:2013–2016
2. Cai L, Friedman N, Xie XS (2006) Stochastic protein expression in individual cells at the single molecule level. Nature 440:358–362
3. Carlo DD, Lee LP (2006) Dynamic single-cell analysis for quantitative biology. Anal Chem 78:7918–7925
4. Sun Y-P, Zhou B, Lin Y, Wang W, Fernando KS et al (2006) Quantum-sized carbon dots for bright and colorful photoluminescence. J Am Chem Soc 128:7756–7757
5. Yu J, Xiao J, Ren X, Lao K, Xie XS (2006) Probing gene expression in live cells one protein molecule at a time. Science 311:1600–1603
6. Augspurger AE, Stender AS, Han R, Fang N (2014) Detecting plasmon resonance energy transfer with differential interference contrast microscopy. Anal Chem 86:1196–1201
7. Liu GL, Long YT, Choi Y, Kang T, Lee LP (2007) Quantized plasmon quenching dips nanospectroscopy via plasmon resonance energy transfer. Nat Methods 4:1015–1017
8. Nie SM, Emory SR (1997) Probing single molecules and single nanoparticles by surface-enhanced Raman scattering. Science 275:1102–1106
9. Futamata M, Maruyama Y, Ishikawa M (2004) Adsorbed sites of individual molecules on Ag nanoparticles in single molecules sensitivity surface-enhanced Raman scattering. J Phys Chem B 108:13119–13127
10. Das P, Metiu H (1985) Enhancement of molecular fluorescence and photochemistry by small metal particles. J Phys Chem 89:4680–4687

11. Andrew P, Barnes W (2004) Energy transfer across a metal film mediated by surface plasmon polaritons. Science 306:1002–1005
12. Boussaad S, Pean J, Tao N (2000) High-resolution multiwavelength surface plasmon resonance spectroscopy for probing conformational and electronic changes in redox proteins. Anal Chem 72:222–226
13. Haes AJ, Zou S, Zhao J, Schatz GC, Van Duyne RP (2006) Localized surface plasmon resonance spectroscopy near molecular resonances. J Am Chem Soc 128:10905–10914
14. Choi Y, Kang T, Lee LP (2009) Plasmon resonance energy transfer (PRET)-based molecular imaging of cytochrome c in living cells. Nano Lett 9:85–90
15. Choi Y, Park Y, Kang T, Lee LP (2009) Selective and sensitive detection of metal ions by plasmonic resonance energy transfer-based nanospectroscopy. Nat Nanotechnol 4:742–746

Chapter 7
Electron Transfer on Plasmonics Surface

Abstract Plasmonic nanoparticles have abundant free electrons on the surface, which make them good electron acceptors and donors and extend their applications in photovoltaic conversion and catalysis, especially providing a platform for the investigation of electron transfer on single nanoparticles. In this chapter, we introduce the influence of electron density on plasmon resonance and the combination of dark-field microscopy with electrochemistry. Also, the charge separations between semiconductor and metal nanoparticles are highlighted.

Keywords Electron density • Charge separation • Electron transfer • Electrochemistry • Photoelectric conversion

7.1 Electron Charging on Plasmonics Surface

The surface plasmon bands of gold nanoparticles are dependent on the surface electron density, as described by Eq. (7.1) [1]. Here, $\Delta\lambda_{max}$ is the wavelength peak shift, λ is the wavelength of incident light, ε_m is the dielectric constant of the surrounding environment, ε is the dielectric constant of the GNP, N is the electron density of GNPs, and L is the shape factor. It has been confirmed that as the plasmonic surface electron density increased, the plasmon resonance energy would increase following the LSPR band blueshift.

$$\Delta\lambda_{max} = -\frac{\Delta N}{2N}\lambda\sqrt{\varepsilon + \left(\frac{1-L}{L}\right)\varepsilon_m} \qquad (7.1)$$

Y.-T. Long and C. Jing, *Localized Surface Plasmon Resonance Based Nanobiosensors*,
SpringerBriefs in Molecular Science, DOI: 10.1007/978-3-642-54795-9_7,
© The Author(s) 2014

Based on this specific property of plasmonic nanoparticles, the reaction rates of catalytic processes on single nanocrystals were determined. For a nanoparticle-catalysed reaction, two redox reactions occur on the surface of the catalyst, as shown in Eqs. (7.2) and (7.3):

$$A^- = A + eNC^- \qquad (7.2)$$

$$B + eNC^- = B^- \qquad (7.3)$$

The donor molecules first transfer electrons to the catalyst nanoparticles, and the acceptor molecules subsequently obtain electrons from the catalyst. If the rate of these processes is faster than the direct electron transfer from A^- to B, the redox reaction will be catalysed by the nanoparticles. Because the scattering of plasmonic nanoparticles depends on their surface electron density, the electron transfer rate of the catalytic reaction can be monitored by the shifts in the scattering spectra of the particles.

After ascorbic acid was added to the surface of gold nanoparticles, their scattering spectra exhibited a blueshift, as shown in Fig. 7.1. The increased scattering energy was caused by the injection of electrons during the catalytic oxidation of ascorbic acid on the surface of the gold nanoparticles. Subsequently, the scattering spectra of the nanoprobe shifted to its initial position in one hour because of the loss of the surface electrons to soluble oxygen. Moreover, the scattering changes due to electron injection and loss were observed for nanoparticles of various shapes. Nanocrystals with smaller shape factors were found to exhibit larger shifts in their scattering peaks, and gold decahedra were observed to exhibit the best catalytic activity. In this work, approximately 4,600 electrons per second could be measured, and only 65 O_2 molecules per second were involved in the chemical reaction. In addition, other similar catalytic reactions on plasmonic surfaces were detected, including electron injection by $NaBH_4$, which acts as a stronger electron donor than ascorbic acid [2]. In the presence of ascorbic acid, the surface plasmon band redshifted due to increased faceting and developed {111} faces, as shown in Fig. 7.2. However, in the presence of $NaBH_4$, the nanorods were fragmented due to the excess electrons injected by $NaBH_4$ onto the particle surface. This phenomenon can be explained by the Rayleigh equation, as described in Eq. (7.4):

$$Q = 8\pi \left(\varepsilon_0 \gamma a^3 \right)^{1/2} \qquad (7.4)$$

where Q is the charge on the sphere, a is the sphere radius, γ is the surface tension, and ε_0 is the permittivity of the surroundings. When Q is greater than the threshold value, the nanoparticle will be unstable, leading to the fragmentation of the nanorods into small spheres.

Mulvany then constructed an integrated instrument that combines dark-field microscopy and an electrochemistry workstation equipped with scattering spectroscopy to investigate the resonance band shifts in response to electrochemical

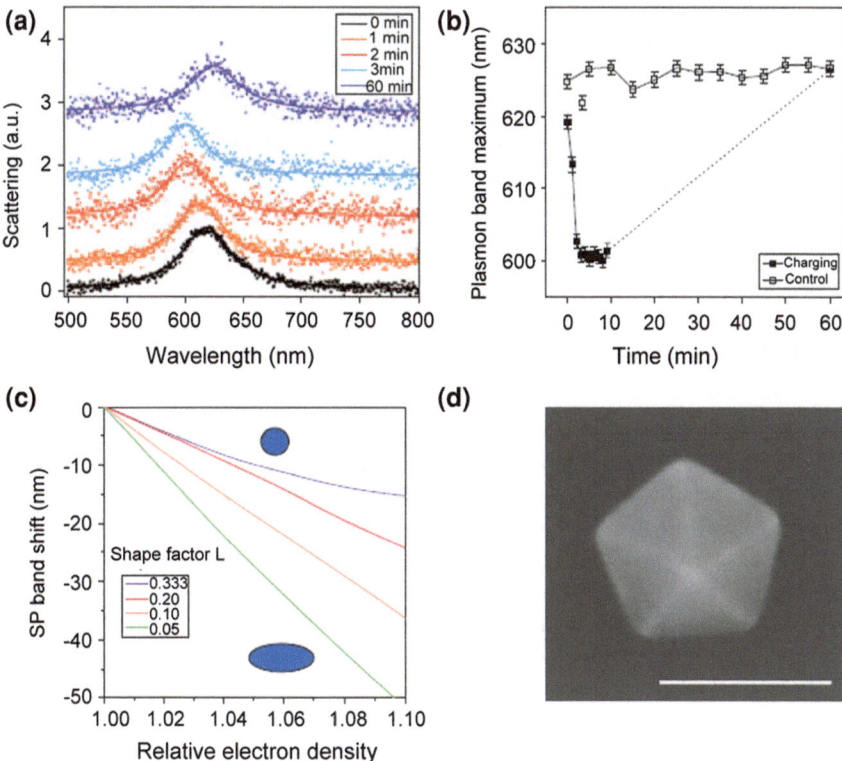

Fig. 7.1 Gold-catalysed oxidation of ascorbic acid. **a** Scattering spectra of the decahedron shown in d before and at 1, 2, 3, and 60 min after electron injection by ascorbate ions. **b** Spectral shift as a function of time for the catalysis reaction and for the control experiment. The error bars represent the error in determining the peak position from the Lorentzian fitting procedure. **c** Plot of the sensitivity of the SP resonance to changes in electron density for different shaped gold nanocrystals. The medium relative permittivity is $\varepsilon_m = 2.025$, and L is the depolarisation factor, which describes the nanocrystal morphology. **d** SEM image of the gold decahedron used in the catalysis experiments. Scale bar, 100 nm. Reprinted with permission from Ref. [1]. Copyright (2008) Nature Publishing Group

charging [3]. The results demonstrated that, under applied potentials of -0.2 to -1.4 V, the light scattering of single GNPs undergoes clear and reversible blueshifts, in good agreement with theoretical calculations. Furthermore, the influence of the shape factor of the GNPs on the scattering shift during charge injection was investigated. As the shape factor L decreased, the GNPs were more sensitive to charge changes on the surface, resulting in more obvious peak shifts, as shown in Fig. 7.3. This method provides a new way to monitor redox reactions and electron transfer on single nanoparticle surfaces, with important applications in catalysis

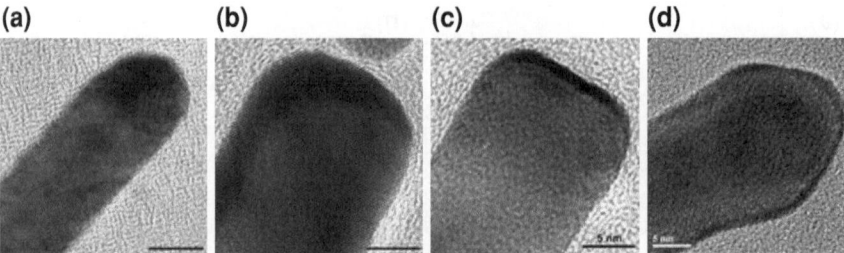

Fig. 7.2 HRTEM images of the end caps of nanorods of initial aspect ratio 4 (**a**) before and after addition of (**b, c**) 1 mM and (**d**) 50 mM ascorbic acid. Scale bar is 10 nm in the first image and 5 nm in (**b–d**). Reprinted with permission from Ref. [2]. Copyright (2007) American Chemical Society

and electrochemistry. Subsequently, Pan developed a novel method for real-time monitoring of the deposition of single silver nanoparticles according to this system. Utilising the relationship between the plasmon resonance intensity and particle size on the basis of Mie theory, the entire reduced silver atoms on electrode were calculated. Thus, corresponding to the whole electrochemical voltammetric curves, it is capable to calculate the electrochemical current on single plasmonic nanoparticles, which is extremely difficult in traditional electrochemistry. This method may offer a novel approach to express the understanding of electrochemical process [4].

By combining dark-field microscopy and electrochemistry, Krenn and co-workers investigated the conductive surface effect on plasmon resonance [5]. In this system, gold nanospheroid array was firstly fabricated on ITO substrate with length of 140 nm and height of 40 nm by electron beam lithography (EBL) as shown in Fig. 7.4a. Then, a submicrometer-thick polyaniline (PANI) layer was deposited on ITO and gold under galvanostatic conditions. The conductivity of PANI film could be modulated through changing its redox state by electrochemistry from oxidised conductive state to reduced nonconductive state with ultrafast switching (in the microsecond range). The thickness of PANI films (20–100 nm) was controlled via the applied deposition potential and time. It was found that the plasmon resonance band was changed following the redox state of PANI by applying electrochemical potential as shown in Fig. 7.4b. The spectral shift was attributed to both the dielectric constant changing and charge carrier density of PANI. In addition, the thickness of PANI also has obvious influence on LSPR spectra as shown in Fig. 7.4b.

Recently, this technique was used to investigate the voltage-induced adsorbate damping of single-gold-nanorod plasmon resonance scattering spectra via cyclic voltammetry (CV) [6]. Cyclic voltammograms were recorded from −1.0 to 2.3 V, as shown in Fig. 7.5. In the potential range of −1.0 to 0.3 V, double-layer charging created an interlayer on the surface of the gold nanorod that led to slight energy loss without resonance damping. In region ii, the reversible adsorption of water molecules and ions in the electrolyte can provide high-energy electronic trap states, resulting in plasmon scattering decay. At potentials greater than 1.0 V, the dominant process is the oxidation of gold nanorods, which causes redshifts in the plasmon resonance but no additional damping. In addition, different electrolyte and potential

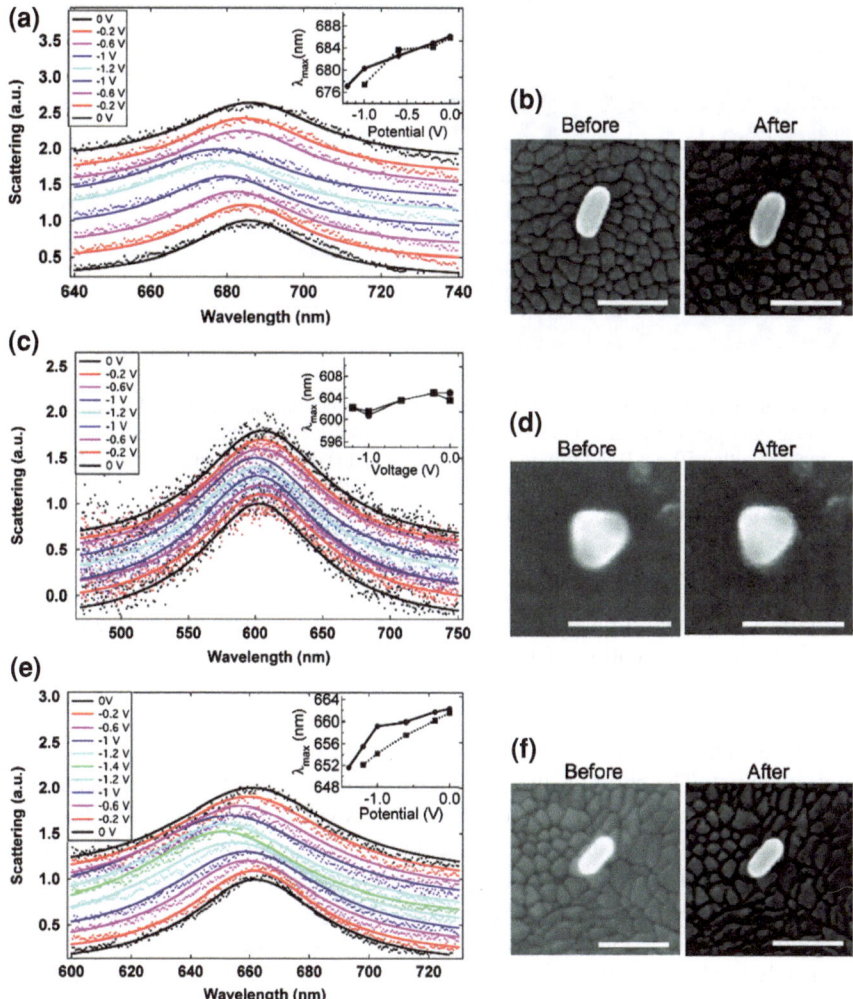

Fig. 7.3 Normalised scattering spectra of gold nanoparticles at potentials varying from 0 V to (**a** and **c**) −1.2 V and (**e**) −1.4 V and back to 0 V. Full lines are Lorentzian fits to the spectra, which are offset for clarity. Insets are plots of λ_{max} versus potential. **b**, **d**, and **f** are SEM images of the particles in (**a**, **c**, **e**), respectively, before and after the charging; the granular structure is the ITO. Scale bars = 100 nm. Reprinted with permission from Ref. [3]. Copyright (2009) American Chemical Society

ranges were considered in the scattering damping. This research improved our understanding of electrochemistry and optoelectronic switching.

Dark-field microscopy provides new insights into electrochemical processes by allowing reactions to be optically observed on single nanoparticles. Harnessing this concept, the electrocatalytic oxidation of H_2O_2 on single gold nanorods was monitored, as shown in Fig. 7.6. In the absence of H_2O_2, the scattering spectra of

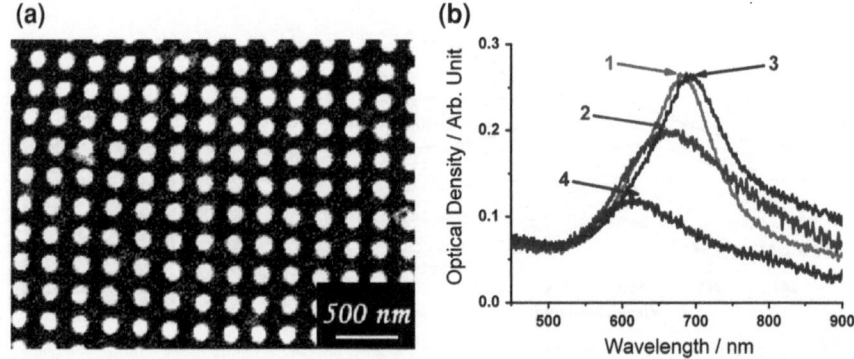

Fig. 7.4 a SEM image of gold nanoparticle array. **b** Extinction spectra under X-polarised light of oblate gold particle grating (*1*) overcoated with a 50-nm PANI film in its reduced state, (*2*) overcoated with a 50-nm PANI film in its oxidised state, (*3*) overcoated with a 100-nm PANI film in its reduced state, and (*4*) overcoated with a 100-nm PANI film in its oxidised state. Adapted with permission from Ref. [5]. Copyright (2008) American Chemical Society

the GNRs indicated a clear redshift in response to the applied positive potential that induced the oxidation of the surface gold atoms. However, during the electrocatalytic oxidation of H_2O_2, the particles underwent a redox reaction and their spectra showed a diminished redshift compared with that in the absence of H_2O_2. This result indicated that H_2O_2 reduced the gold oxide produced by the applied high potential. The shifts in the spectra of the particles indicated the occurrence of a catalytic reaction, and different particles provided different results. Based on this method, individual nanoparticles could be screened for catalytic efficiency. In addition, the reaction was investigated in the presence of chloride ions, which revealed that chloride ions strongly interact with gold, resulting in the formation of a gold–chloride complex. Interestingly, the gold–chloride complex can block the catalytic ability of the gold nanoparticles, resulting in no obvious changes in the spectra during electrochemical scanning before and after treatment with H_2O_2. This type of spectral information allowed the investigation of the reaction mechanism and may improve the exploration of electrochemistry and catalysis.

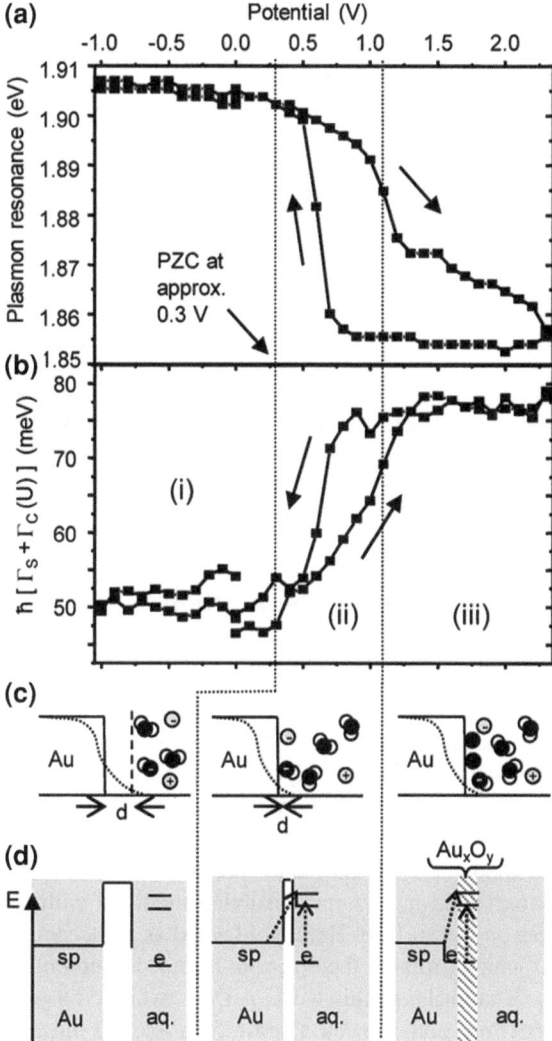

Fig. 7.5 Plasmon resonance peak energy (**a**) and damping $\hbar$ ($\Gamma_S + \Gamma_C$ (U)) (**b**) versus applied potential on the full potential range from -1 to $+2.3$ V. **c** Sketch of the positive jellium background in the AuNR (*solid black line*) and the sp-electron density (*dotted line*). Three different regions can be associated: (*i*) charging of double-layer capacitance leads to a spectral shift but no additional damping. An interlayer of thickness d is formed where the electron spill-out repels both the solvent molecules and dissolved ions (*left sketch* in **c**). No chemical damping takes place via the solvent or dissolved ions (*left sketch* in **d**). In region (*ii*), potentials positive of the point of zero charge (PZC) cause a rapid nonlinear redshift (*anodic scan* in **a**) and substantial additional damping [region (*ii*) in panel **b**]. The sp-electron spill-out retracts (reducing d), and solvent molecules and anions adsorb (*centre sketch* in **c**). Chemical surface damping of the nanoparticle plasmon resonance (NPPR) becomes possible by the excitation of either sp-electrons or adsorbate electrons into empty adsorbate states (*dotted arrows* in the *centre sketch* of **d**). (*iii*) At potentials above 1.1 V, Au oxidation leads to no additional damping (**b**) but some further spectral redshift (**a**) due to the trapping of sp-electrons at the oxide (*right sketches* in **c**, **d**). Reprinted with permission from Ref. [6]. Copyright (2012) American Chemical Society

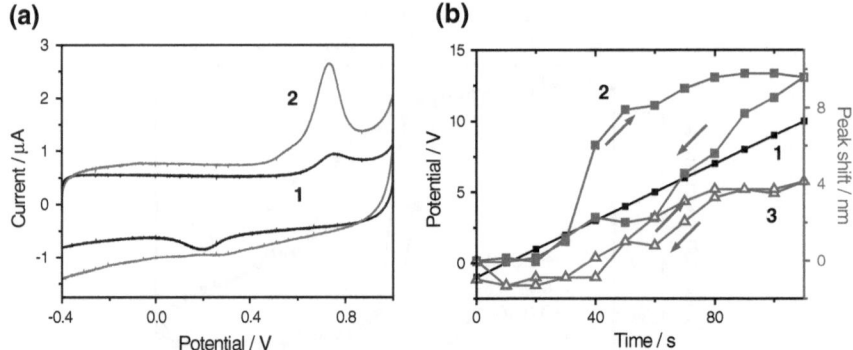

Fig. 7.6 a CVs of GNRs in 0.10 M KNO_3 solution (*1*), and 0.10 M KCl solution (*2*), scan rate 100 mV s^{-1}, Pt quasi-reference electrode. **b** Scattering spectra $\Delta\lambda_{max}$ of single GNR in the presence (*2*) and absence (*3*) of 1.00 mM H_2O_2 in 0.10 M KNO_3 solution under the applied potential (*1*). 538. WE-Heraeus-Seminar on "Light at the Nanotip: Scanning Near-field Optical Microscopy and Spectroscopy", Bad Honnef, Germany, 2013

7.2 Plasmon-Induced Charge Separation

The electron transfer between gold nanoparticles and semiconductor is a hot topic in photocatalysis and solar cells. Notably, Cu_2O, a p-type semiconductor with band gap of 2.0–2.3 eV, as photoexcited semiconductor metal oxides has been widely applied in the degradation of organic dyes via the formation of hydroxyl radicals resulting from the electron–hole separation [7]. That is, the excitation-formed holes can oxidise water into hydroxyl radicals as shown in Fig. 7.7, which will attack the double bonds of dyes and induce the destruction and bleaching of dyes. El-Sayed synthesised gold nanoframes (AuNFs) modified with Cu_2O shells with different thickness on the surface. Here, gold acted as an accepter to receive electrons from Cu_2O which promote the electron–hole separation of Cu_2O, leading to more production of radicals. In this work, AuNFs exhibited a plasmon resonance peak at about 1030 nm and showed a redshift to ~1,200 nm as the shell thickness increased. From both the rate of photodegradation of organic dye (methylene blue) and decay rate of the exciton, it was confirmed that as the thickness of Cu_2O increased, the nanoparticles had higher electron–hole separation efficiency and the Cu_2O–Au was more active than bare Cu_2O, proving the function of gold inside.

Gold nanoparticles also enable the enhancement of charge separation in TiO_2, which provide a promising photovoltaic conversion system. Tachiya confirmed that after the irradiation of visible light, the electrons would transfer from gold nanoparticles to TiO_2 ultrafast (<240 fs) with a yield of 40 %, inducing the charge separation [8]. Tatsuma and Tian investigated the charge separation of Au–TiO_2 mechanism and optimised the electron donor, as well as the application in photocatalytical oxidisation of ethanol and methanol [9, 10]. Figure 7.8 illustrates that the electrons transferred from the donor to excited gold nanoparticles and then

(a)   **(b)**

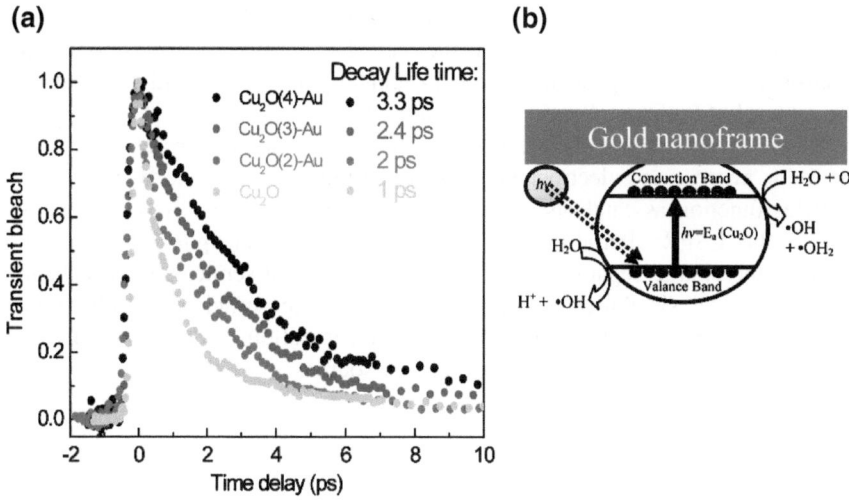

Fig. 7.7 a The transient bleach intensity as a function of delay time for pump wavelength at 400 nm and probe at 460 nm. The decay curves result from the exciton relaxation at electron–hole recombination and transfer to the gold and to the defects at its interface. Faster decay is observed for thinner Cu_2O samples or smaller Cu_2O–Au separation. **b** Schematic diagram summarising the excitation of Cu_2O by visible light and the mechanism of radical formation. Reprinted with permission from Ref. [7]. Copyright (2011) American Chemical Society

Fig. 7.8 Proposed mechanism for the photoelectrochemistry. Charges are separated at a visible-light-irradiated gold nanoparticle–TiO_2 system. Reprinted with permission from Ref. [9]. Copyright (2005) American Chemical Society

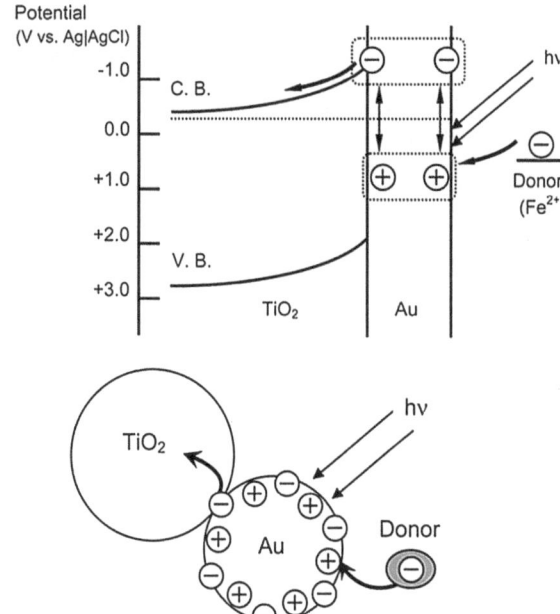

injected to TiO_2. This $Au-TiO_2$ can be applied as visible-light-sensitive photocatalysts in a new kind of photovoltaic fuel cell. Furthermore, the plasmonic TiO_2 electron transfer is dependent on the incident light which could modulate the energy level of both metal nanoparticles and TiO_2.

Therefore, monitoring the electron transfer on plasmonics surface could reveal the redox process and electrochemical reaction kinetics. Particularly, when nanoparticles function as catalysts, the catalysis mechanism is readily to be obtained due to the LSPR alternation including intensity, wavelength, and linewidth. Notably, the charge separation between plasmonics and semiconductor enhances the optical catalysis ability and photovoltaic conversion efficiency of photosensitiser. Development of plasmon-induced charge separation materials is a prospective direction in the exploration of new energy sources.

References

1. Novo C, Funston AM, Mulvaney P (2008) Direct observation of chemical reactions on single gold nanocrystals using surface plasmon spectroscopy. Nat Nanotechnol 3:598–602
2. Novo C, Mulvaney P (2007) Charge-induced Rayleigh instabilities in small gold rods. Nano Lett 7:520–524
3. Novo C, Funston AM, Gooding AK, Mulvaney P (2009) Electrochemical charging of single gold nanorods. J Am Chem Soc 131:14664–14666
4. Hill CM, Pan S (2013) A dark-field scattering spectroelectrochemical technique for tracking the electrodeposition of single silver nanoparticles. J Am Chem Soc 135:17250–17253
5. Leroux Y, Lacroix JC, Fave C, Trippe G, Félidj N, Aubard J et al (2008) Tunable electrochemical switch of the optical properties of metallic nanoparticles. ACS Nano 2:728–732
6. Dondapati S, Ludemann M, Müller R, Schwieger S, Schwemer A, Händel B et al (2012) Voltage-induced adsorbate damping of single gold nanorod plasmons in aqueous solution. Nano Lett 12:1247–1252
7. Mahmoud MA, Qian W, El-Sayed MA (2011) Following charge separation on the nanoscale in Cu_2O-Au nanoframe hollow nanoparticles. Nano Lett 11:3285–3289
8. Furube A, Du L, Hara K, Katoh R, Tachiya M (2007) Ultrafast plasmon-induced electron transfer from gold nanodots into TiO_2 nanoparticles. J Am Chem Soc 129:14852–14853
9. Tian Y, Tatsuma T (2005) Mechanisms and applications of plasmon-induced charge separation at TiO_2 films loaded with gold nanoparticles. J Am Chem Soc 127:7632–7637
10. Takahashi Y, Tatsuma T (2010) Electrodeposition of thermally stable gold and silver nanoparticle ensembles through a thin alumina nanomask. Nanoscale 2:1494–1499

Chapter 8
Plasmonic Nanoparticles in Cell Imaging and Photothermal Therapy

Abstract Plasmonic nanoparticles have been widely used in biosensors, especially in cell imaging, because of their excellent biocompatibility and optical properties. Compared with fluorescent dyes or quantum dots, the scattering of plasmonic nanoparticles is more stable and does not undergo photobleaching or photoblinking. Furthermore, the toxicity of gold nanoparticles is extremely low, especially after the biomodification. In this chapter, we mainly introduce the applications of plasmonics in cell imaging and thermal therapy.

Keywords Cell imaging • Phototherapy • Toxicity of nanoparticles • Multifunctional probes • In vivo detection

8.1 Toxicity Testing of Nanoparticles

For cell imaging or in vivo research, the toxicity of nanoprobes is a critical factor that must be assessed [1–4]. Nanoparticles with different sizes and surface modifications must undergo biocompatibility testing prior to being used in living cells. The K562 leukemia cell line was used to test the toxicity of gold nanoparticles by exposing the cells to the nanoparticles for three days as shown in Fig. 8.1 [5]. Cell viability was observed using an MTT assay. The results demonstrated that 18-nm gold nanoparticles with citrate and biotin surface modifiers are nontoxic at concentrations up to 250 μM (gold atoms). Nanoparticles modified with glucose, cysteine or other reducing reagents exhibited no toxicity at concentrations up to 25 μM. In contrast, the as-prepared 18-nm nanoparticles modified with CTAB exhibited significant toxicity towards cells, and CTAB solution exhibited similar toxicity. Thus, gold nanoparticles themselves are nontoxic to cells. However, in this work, only gold spheres were investigated. Therefore, other shapes and compositions as well as the long-term influence of plasmonic particles on the cell growth and uptake processes need to be tested.

Y.-T. Long and C. Jing, *Localized Surface Plasmon Resonance Based Nanobiosensors*,
SpringerBriefs in Molecular Science, DOI: 10.1007/978-3-642-54795-9_8,
© The Author(s) 2014

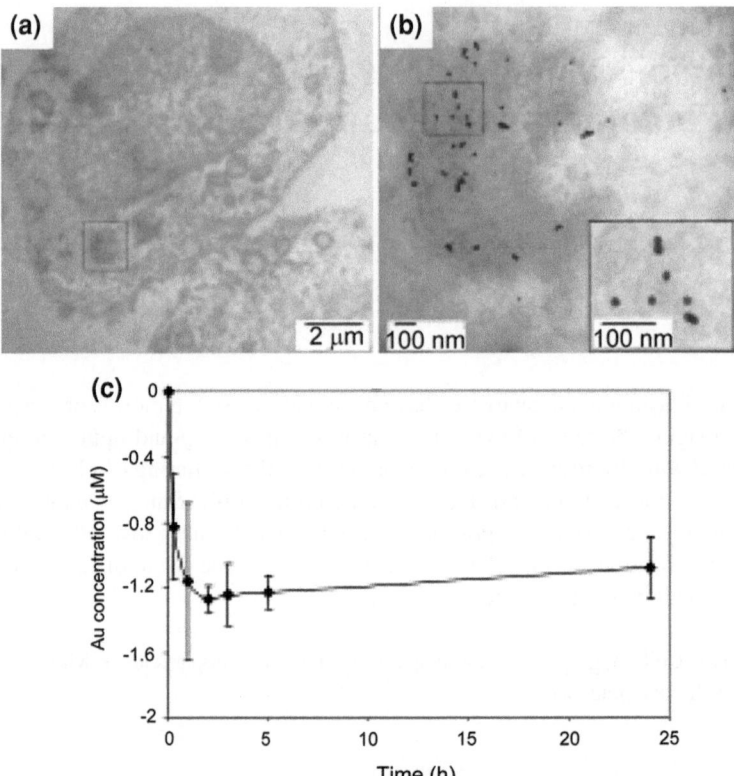

Fig. 8.1 Electron micrographs at different magnifications of a cell containing nanoparticles. Cells were exposed to nanoparticles for 24 h, fixed with osmium tetroxide, sectioned, and visualised with a Hitachi H-8000 electron microscope. **a** Image at 8,000× magnification of a representative cell with nanoparticles subcellularly localized. The small box represents the area magnified in (**b**). **b** Image at 60,000× magnification of gold nanoparticles within cells. The inset is a 150,000× magnification of the gold nanoparticles. **c** Visible spectroscopy plot (measured at 526 nm) of the concentration of gold nanoparticles in cell culture media following incubation with the cells (data is compared to the initial concentration). The media were exposed to 18-nm citrate-capped nanoparticles for the times shown. Following exposure, the cells were removed from the media by centrifugation at 300 g. Cells were grown in cell culture media lacking phenol red for the absorbance experiments. Reprinted with permission from Ref. [5]. Copyright 2005 WILEY-VCH Verlag GmbH & Co. KGaA, Weinheim

Recently, a new method based on dark-field microscopy and transmission electron microscopy was used to investigate the uptake process and toxicity towards cells of gold nanoparticles with different shapes, surface charges, and stabilising agents [6]. The toxicity of gold nanoparticles is related to their internalisation into cells. Thus, the observation of the assembly, the distribution, and uptake of gold nanoparticles can provide versatile information about the interactions between nanoparticles and cells. This method enables the quantification of the number of particle aggregates in cells and the number of particles per aggregate, as shown in Fig. 8.2. The number of

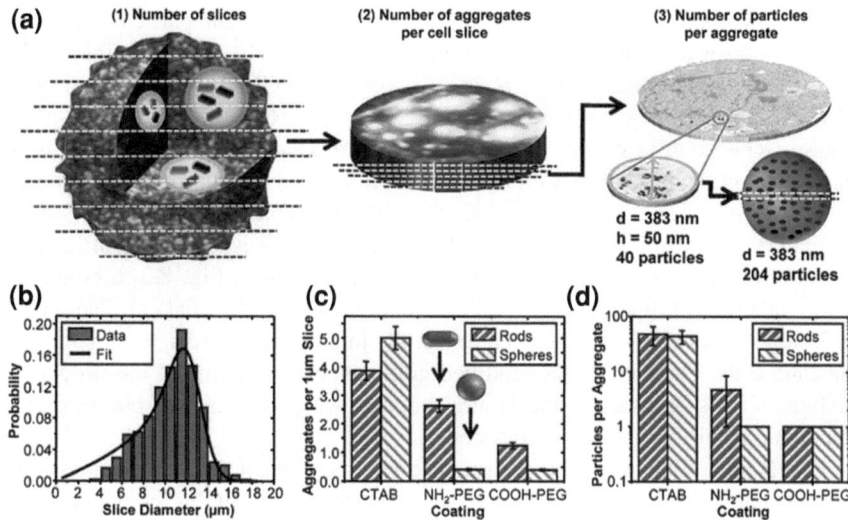

Fig. 8.2 a Scheme to illustrate the determination of the total number of particles per cell N as the product of (1) the number of slices per cell N_{slices}, (2) the number of aggregates per cell slice $N_{aggregates}$, and (3) the number of particles per aggregate $N_{particles}$ yielding the formula $N = N_{slices} \times N_{aggregates} \times N_{particles}$. (1) N_{slices} results from the cell diameter divided by the slice thickness N_{slices} = diameter/thickness. The mean cell diameter is determined from DF images. (2) $N_{aggregates}$ is derived from the number of bright spots within the cell area observed by DF. (3) $N_{particles}$ is computed from the particles in one aggregate as counted on TEM images and corrected for the three-dimensional volume of the aggregate. **b** Histogram of the diameters of 1-μm cell slices measured in optical microscopy and fitted numerically assuming spherical cells with a Gaussian size distribution cut at random positions in 1-μm slices. The fit yields a mean cell diameter of (12.5 ± 0.6) μm, that is $N_{slices} = (12.5 \pm 0.6)$ slices per cell of 1 μm thickness. **c** $N_{aggregates}$ and its uncertainty determined from DF images. **d** $N_{particles}$ and its uncertainty determined from TEM images. If no particle was found, the number of particles per aggregate was set to one in order to assign the lowest meaningful value. Reprinted with permission from Ref. [6]. Copyright 2012 WILEY-VCH Verlag GmbH & Co. KGaA, Weinheim

internalised nanoparticles depends primarily on the nature of the surface modification rather than the particle shape. CTAB-coated nanoparticles exhibit strong uptake in cells and are located in multivesicular bodies or late endosomes. In contrast, neutral NH_2-PEG particles and slightly negative COOH-PEG particles were taken up through nonspecific endocytosis or macropinocytosis, which generated large protrusions of the cell membrane around the particles.

8.2 Functional Plasmonic Nanoparticles in Cell Imaging

Gold nanoparticles have considerable potential in cancer cell imaging because of their excellent biocompatibility, nontoxicity, and stability. The targeted delivery of nanoparticles to solid tumours is one of the most important challenges in cancer imaging, nanomedicine, and therapy [7]. However, gold nanoparticles alone are nonspecific towards malignant and nonmalignant cells. To overcome this limitation, the antiepidermal growth factor receptor (anti-EGFR) was used to modify GNPs to enhance their targeting of cancer cells [8]. EGFR is a receptor tyrosine kinase (RTK) that is widely distributed on the surfaces of epithelial cells, fibroblasts, and gliocytes and plays a vital role in fundamental cellular processes, such as DNA replication and cell division. The EGFR internalisation process is a key regulatory pathway for monitoring cell behaviour. EGFR is an important factor in the development of new cancer therapeutics. As shown in Fig. 8.3, a nonmalignant epithelial cell line (HaCaT) and two malignant oral epithelial cell lines (HOC 313 clone 8 and HSC-3) were used to test the interaction between gold nanoparticles and cells. The anti-EGFR-modified gold nanoparticles bind to the surfaces of malignant cells specifically and homogeneously with an affinity of 600 % greater than their affinity for noncancerous cells.

EGFR signalling was then measured after internalisation in early endosomes [9]. The organisation, aggregation, and sequestration of EGFR could be easily imaged at the nanoscale by taking advantage of the coupling of nanoparticles. Sokolov utilised plasmonic scattering to monitor EGFR in living A431 cells. Real-time images of living cells incubated with anti-EGFR-modified nanoparticles revealed distinct colour changes from green to yellow to orange, as displayed in Fig. 8.4b. This result confirms the trafficking dynamics of EGFR shown in Fig. 8.4a. Initially, the EGFR molecules adsorbed onto the surfaces of the cells; they then gradually dimerised and aggregated in the membrane, which exhibited a green colour similar to that of the dispersed nanoparticles. Approximately 5–10 min after endocytosis, early endosomes formed and entered the cell plasma, causing the slight coupling of the plasmonic particles and producing red shifts in their scattering spectra. These early endosomes could either recycle to the cell membrane or proceed to form late endosomes and multivesicular bodies (MVB) within 20–60 min, which led to the intense coupling of the nanoparticles. Receptor trafficking resulted in a gradual plasmonic spectral red shift of more than 100 nm over 60 min. This continuous change was arrested by a potent EGFR inhibitor, AG1478, as shown in Fig. 8.4c, which interfered with EGFR transphosphorylation and internalisation. These results revealed that plasmonic particle coupling occurred due to ligands binding to EGFR and their intracellular trafficking in vesicles. This method also offers a creative, near-real-time approach for obtaining distributions of EGFR regulatory states in living cells.

These results suggest that scattering imaging and monitoring of the spectra of specifically modified gold nanoparticles can provide a useful way to investigate cancer cells in vivo and in vitro. For example, gold nanoparticles coated with polyethylene glycol (PEG) were conjugated with an arginine-glycine-aspartic acid peptide (RGD) and a nuclear localization signal (NLS) peptide as shown in Fig. 8.5 [10, 11]. The RGD peptide could specifically target $\alpha_v\beta_6$ integrins on cell surfaces and penetrate the cytoplasm of cancer cells by receptor-mediated endocytosis. NLS, with a sequence of lysine-lysine-lysine-arginine-lysine (KKKRK), is known to be a nuclear-target peptide with the ability to associate

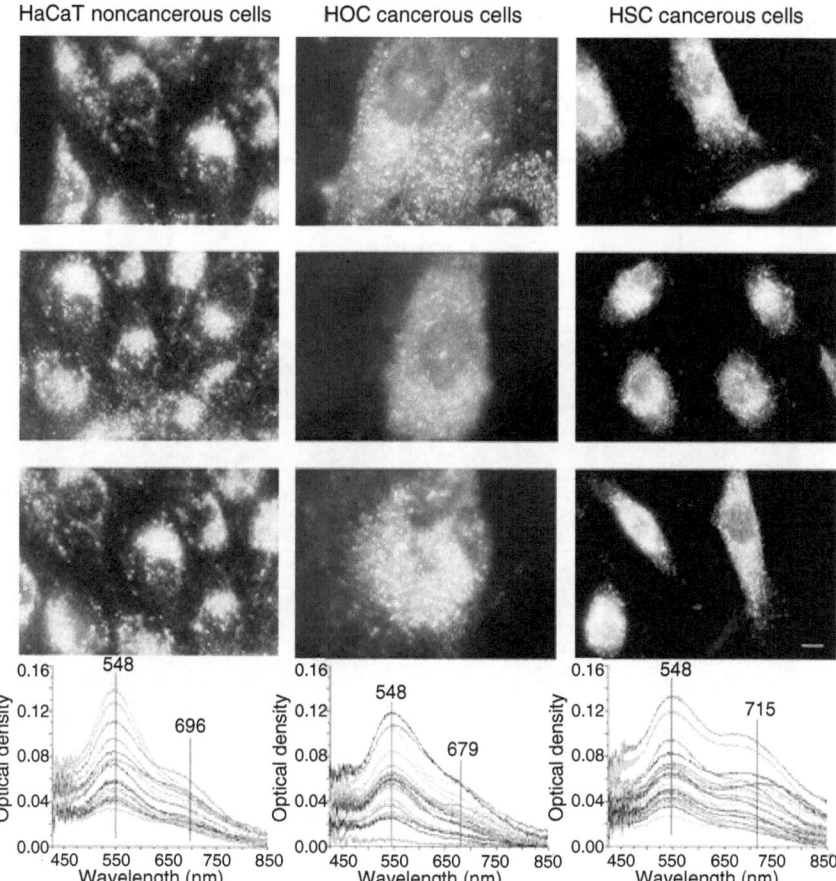

Fig. 8.3 Light scattering images and microabsorption spectra of HaCaT noncancerous cells (*left column*), HOC cancerous cells (*middle column*), and HSC cancerous cells (*right column*) after incubation with unconjugated colloidal gold nanoparticles. Three different images of each kind of cells are shown to test reproducibility. The images show that the particles are inside the cells in the cytoplasm region but do not seem to adsorb strongly on the nuclei of the cells. The absorption spectra were measured for 25 different single cells of each kind. They show that nanoparticles have an SPR absorption maximum around 548 nm, independent of the cell type. The broad, long wavelength tails in the absorption spectra suggest the presence of aggregates. It also shows that no specific difference is observed in either the scattering images or the absorption spectra of the gold nanoparticles in the cancerous and the noncancerous cells. Scale bar: 10 μm for all images. Reprinted with permission from Ref. [8]. Copyright (2005) American Chemical Society

with karyopherins (importins) in the cytoplasm after translocating to the nucleus. The RGD-NLS-modified gold nanoparticles were confirmed to be toxic, inducing DNA damage and apoptotic populations in cancer cells containing cell surface $\alpha_v\beta_6$ integrins such as HSC-3 cell lines. Subsequently, the apoptotic processes of cancer cells were investigated on the basis of plasmonic scattering light imaging

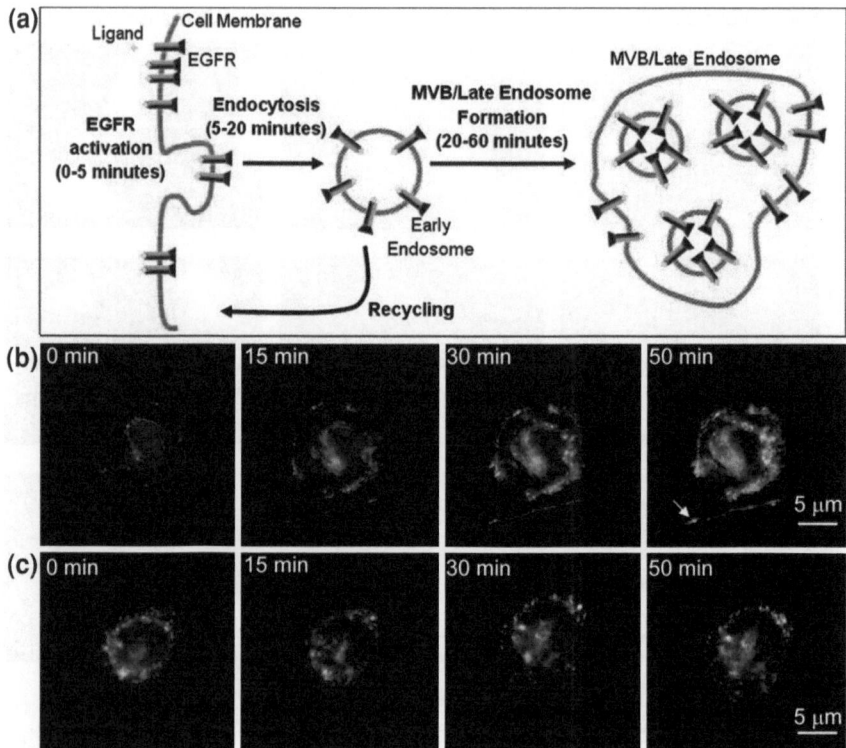

Fig. 8.4 Dynamic imaging of live cells. **a** A schematic of EGFR trafficking upon ligand binding. This process was monitored in real time in live cells labelled with anti-EGFR gold nanoparticles under the microscope. **b** Live A431 cells and **c** live cells pretreated with 10 μM AG1478, which reduces endocytosis. From left to right, images show colour and intensity changes in the gold-nanoparticle optical signal at 0, 15, 30, and 50 min after initial exposure to nanoparticles. Dark-field images of untreated cells (**b**) display changes in colour from blue to yellow orange that are not present in treated cells (**c**). The arrow in panel B, far right, indicates the presence of a filopodium emanating from an adjacent cell. Images were acquired using transmitted dark-field illumination and a long working distance 63×, 0.75 NA objective. Reprinted with permission from Ref. [9]. Copyright (2009) American Chemical Society

by the nuclear-target nanoparticles. After the apoptosis of cells induced by the RGD-NLS nanoparticles, neighbouring cells were observed to act as nonprofessional phagocytes. To avoid a loss of membrane integrity and prevent the release of cytotoxins, the dying cells must be recognised and removed. Thus, the apoptotic cells excrete biological signals, such as annexin I (AnxI) and phosphatidylserine (PS), to attract neighbouring cells to "eat" them. Through these signals, living cells can detect the dying cells and engulf them. In addition, the nanoparticles in the dying cells can generate ROS and permeate neighbouring cell membranes, leading to localized cell death. This work indicates that functionalised plasmonic nanoparticles can be used as contrast agents in cell imaging to detect responses to external stimuli and uptake processes.

0 hours   22 hours

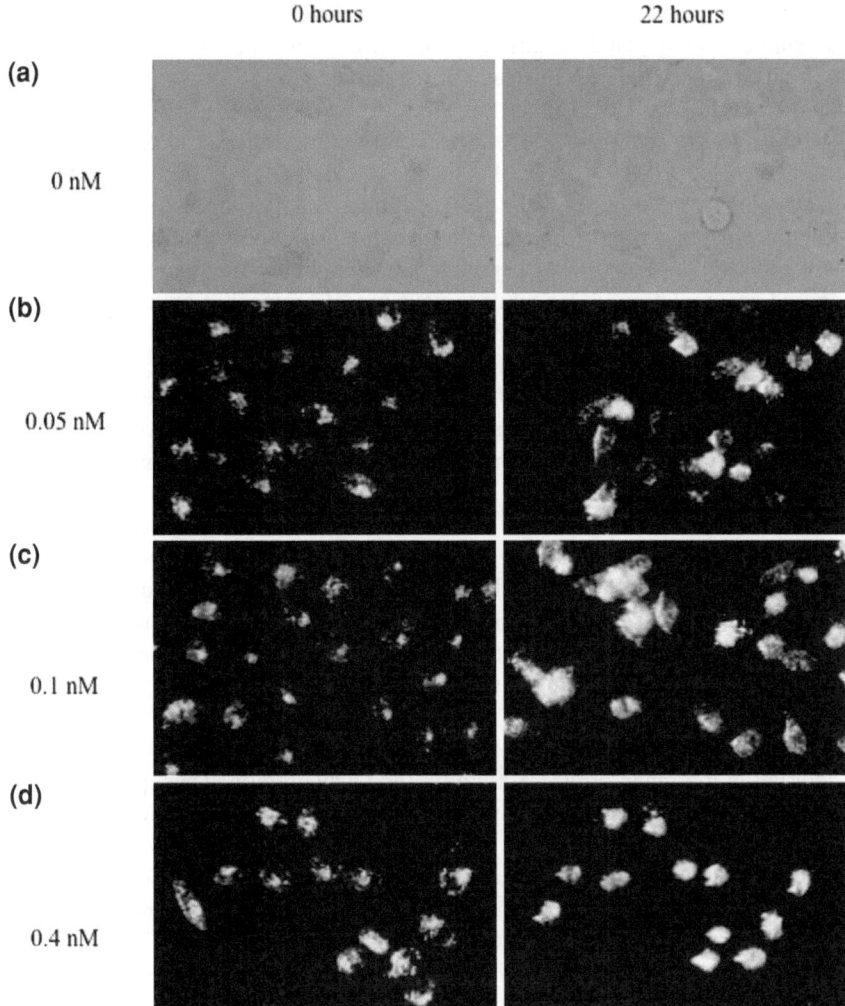

Fig. 8.5 Still images taken at 0 and 22 h of HSC-3 cells incubated with **a** 0, **b** 0.05, **c** 0.1, and **d** 0.4 nM nuclear-targeting AgNPs. Cells treated with 0 nM AgNPs do not show plasmonic light scattering. Images of cells treated with NLS/RGD-AgNPs show that (1) scattering from the AgNPs is heavily localized at the nucleus (2) cellular clustering increases as the AgNP concentration increases from 0.05 to 0.1 nM, and (3) the degree of cellular clustering decreases when cells are treated with 0.4 nM AgNPs. The observed decrease in clustering could be attributed to the increased rate of the observed cell death. The high AgNP density at the cell nucleus, as indicated by the strong scattering (**d**), is the reason for the rapid cell death (see text) for most cells and could impair the ability of these cells to produce or detect intercellular apoptotic signals. Reprinted with permission from Ref. [10]. Copyright (2011) American Chemical Society

   Noble-metal nanoparticles that exhibit LSPR provide a powerful method for monitoring the intracellular interactions of biomolecules for cell imaging and qualitative detection. However, the detailed mechanisms of these processes are difficult to determine quantitatively. Nie reported quantitative tumour uptake observations for nanorods conjugated to different tumour-targeting peptides using elemental mass spectrometry, taking advantage of the fact that gold is not a naturally occurring element in animals [12]. In this work, three different types of ligands were selected: a single-chain variable fragment (ScFv) that recognises EGFR, an amino-terminal fragment (ATF) peptide that recognises the urokinase plasminogen activator receptor (uPAR) and a cyclic RGD peptide that recognises the $\alpha_v\beta_3$ integrin receptor. Quantitative pharmacokinetic and biodistribution results indicated that these three types of targeting peptides only marginally improved the total gold nanorod accumulation in xenograft tumour models compared with the accumulation of nontargeted nanorods. However, unexpectedly, the targeted nanorods enabled the alteration of the intracellular and extracellular nanoparticle distribution, as shown in Fig. 8.6. The c-RGD-modified nanorods exhibit the best intracellular accumulation, and the nontargeted nanorods are difficult to observe in cells. The quantitative data are displayed in Fig. 8.6b and provide direct insights into the uptake efficiency of nanoparticles in living cells.

Active molecular targeting of tumour microenvironments was also confirmed to have no distinct influence on cell uptake during the intravenous injection of nanorods, which revealed that substance transport across the tumour vasculature is a rate-limiting step for nanoparticle delivery and is weakly affected by receptor binding. These results suggest that the most effective route for nanoparticle administration for photothermal cancer therapy is through intratumoural injection rather than intravenous injection.

Nanoparticles with different sizes, shapes, and compositions exhibit varied light scattering in the range of 400–800 nm in the visible region. Thus, plasmonic particles with different scattering colours could serve as multifunctional probes to investigate several metabolism or uptake processes simultaneously in living cells through multiplexed signal detection. Prasad proposed two-colour contrast agents using antibody-conjugated gold nanorods that are red in colour and silver nanospheres that are blue in colour to detect receptor-mediated delivery in human pancreatic cancer cells [13]. The conjugation of the nanoparticles to the biomolecules was based on electrostatic interactions. Gold nanorods coated with CTAB are positively charged. The positively charged surface becomes negatively charged when the nanorods are coated with a PEDT/PSS polyelectrolyte solution to reduce the toxicity of CTAB. To obtain a positively charged surface again, a layer of PDDAC polyelectrolyte was formed on the nanorod surfaces. For the silver nanospheres, the particle surface was negatively charged due to capping by citrate molecules. The negative surface was converted to a positive surface using the previously mentioned method. The modified nanoparticles were subsequently bound to antibodies of anticlaudin 4 (aC4) and antimesothelin (aMe). The corresponding antigen receptors of aC4 and aMe (claudin 4 and mesothelin) are proved to be overexpressed in both primary and metastatic pancreatic cancer. As Fig. 8.7 shows, in dark-field images, the two types of nanoparticles can be clearly distinguished by their red and blue colours, allowing multiplexed imaging with high contrast. This method indicates that plasmonic particles with varying colours can be used as

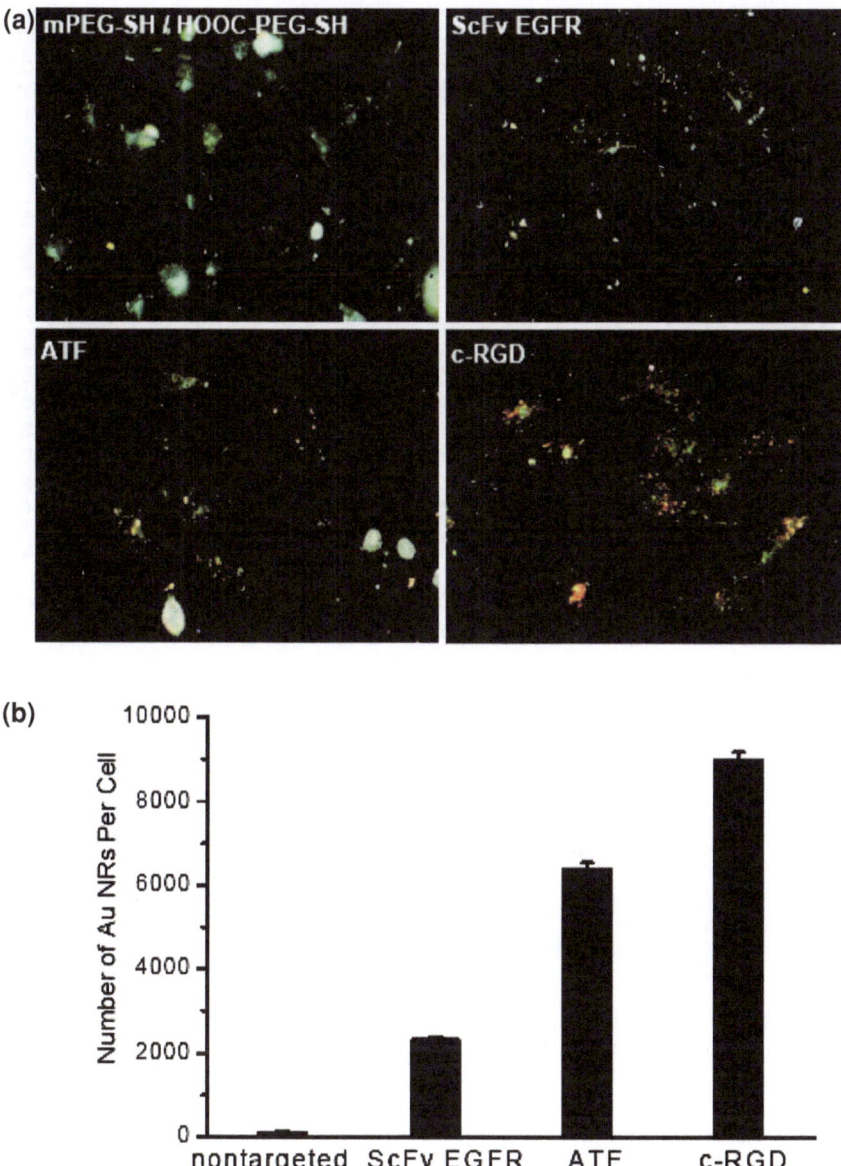

Fig. 8.6 (**a**) Dark-field imaging and (**b**) quantitative Au ICP-MS studies of NR binding to cultured A549 cancer cells. The data show specific binding of peptide-conjugated Au NRs to cultured A549 lung cancer cells, and negligible binding of nontargeted particles to the same tumour cells. Cells were incubated with 1 nM Au NRs for 2 h at 37 °C in the culture medium and were washed with 1 × PBS buffer. After trypsin treatment, approximately one million cells were counted and analysed to minimise statistical errors. Reprinted with permission from Ref. [12]. Copyright (2010) American Chemical Society

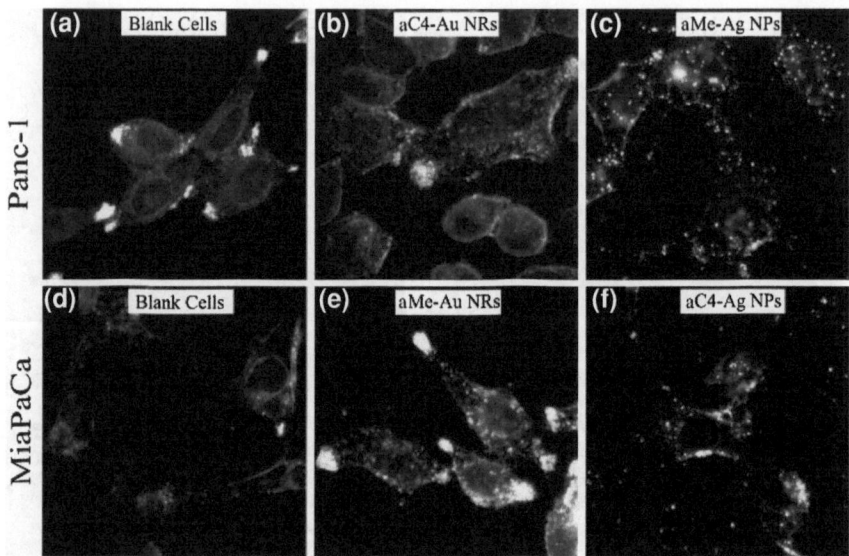

Fig. 8.7 Labelling of Panc-1 and MiaPaCa cells with antibody-conjugated nanoparticles. Reprinted with permission from ref. [13]. Copyright (2009) American Chemical Society

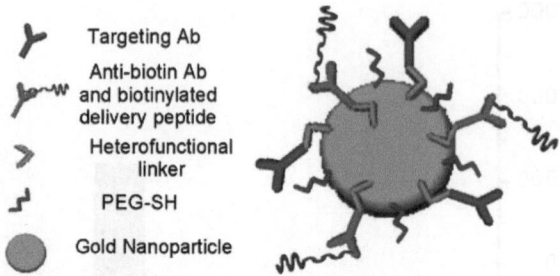

Fig. 8.8 Multifunctional contrast agent. Schematic representation of gold-nanoparticle-based contrast agent with both targeting and delivery components. Reprinted with permission from Ref. [15]. Copyright (2007) American Chemical Society

multiplexed probes for monitoring interactions between proteins or intercellular communication.

Moreover, nanoparticles can be easily modified with various types of molecules to construct multifunctional probes for cell imaging [14]. Sokolov used PEG to fabricate novel four-function nanoparticles that exhibit targeting, endosomal uptake, endosomal release, and improved biocompatibility, as shown in Fig. 8.8 [15]. A monoclonal antiactin antibody specific for the C-terminal end was selected as the targeting molecule for observing actin rearrangement. A monoclonal antibiotin antibody was conjugated to the nanoparticles to allow attachment

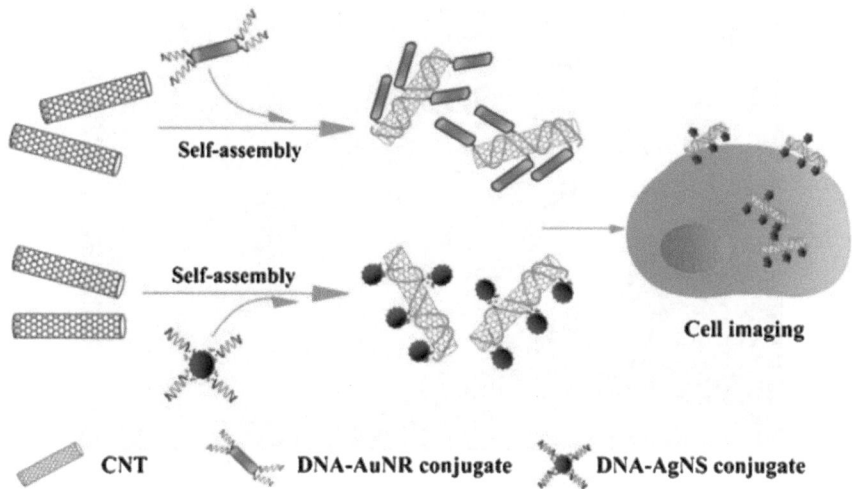

Fig. 8.9 Schematic representation of metal nanoparticle/CNT hybrids fabrication. Reprinted with permission from Ref. [17]. Copyright 2010 Royal Society of Chemistry

of the biotinylated TAT-HA2 peptide, which mediated cytosolic delivery through fusogenic transduction, including both endosomal uptake and release. The anti-actin and anti-biotin antibodies were conjugated to nanoparticles via a heterofunctional linker. These multifunctional contrast agents were incubated with living NIH3T3 fibroblasts and were monitored by both dark-field reflectance and transmission electron microscopy. The multifunctional nanocomplex enabled targeting of various intracellular processes for which cytotoxic labels are not feasible. In addition, nanoparticles with high multiplexing capacity were used to monitor different cell surface markers simultaneously for identity profiling to examine the evolution and distribution of surface markers [16].

A new type of nanomaterial was fabricated for the application of nanoparticles in cell imaging, as shown in Fig. 8.9 [17]. Metal nanoparticle/carbon nanotube hybrids exhibit strong absorption in the NIR region because of the CNTs and exhibit scattering light in the visible region because of the Ag nanoparticles, which makes them excellent agents for cell imaging and cancer therapy.

8.3   In vivo Detections

Plasmonic nanoparticles were used for the real-time imaging of membrane transport in living zebrafish embryos with single-nanoparticle resolution in vivo [18, 19]. In this study, highly purified and stable Ag nanoparticles with various sizes (5–46 nm) were transported into and out of the embryos at three developmental stages including normally developed, deformed, and dead zebrafish to determine their biocompatibility and

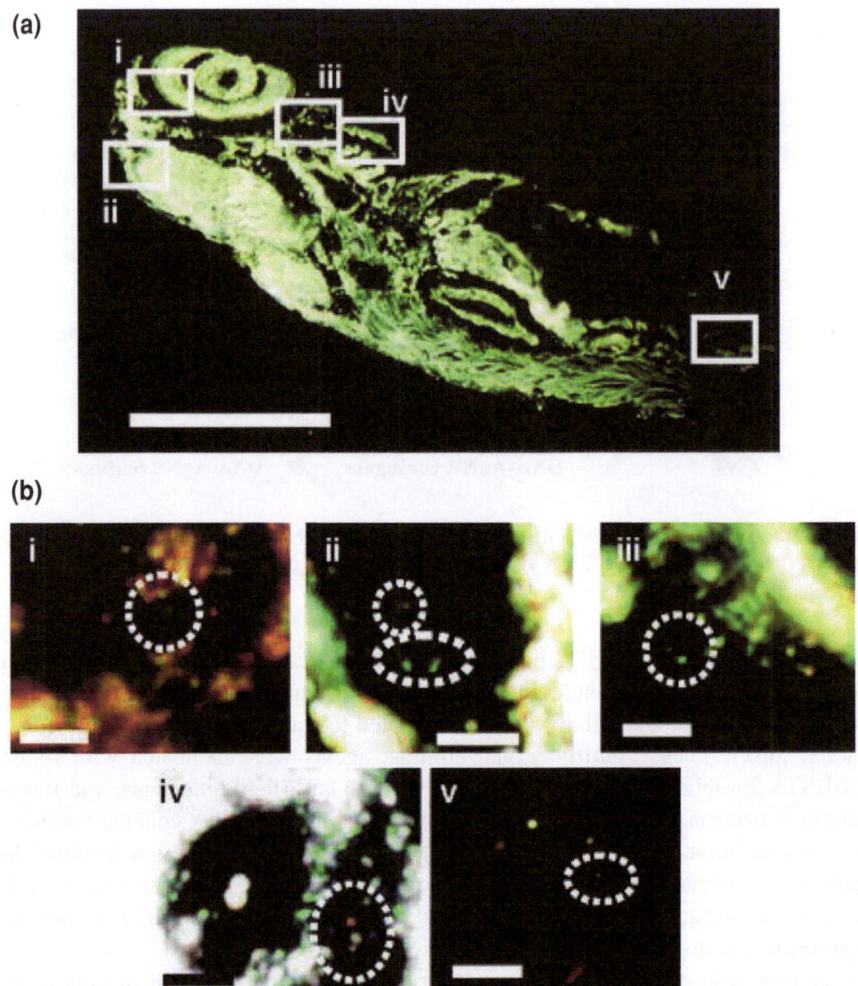

Fig. 8.10 Characterisation of individual Ag nanoparticles embedded inside a fully developed (120 hpf) zebrafish using dark-field SNOMS. **a** Optical image of a fixed, normally developed zebrafish. The rectangles highlight representative areas: (1) retina (2) brain (mesencephalon cavity) (3) heart (4) gill arches, and (5) tail. **b** Zoom-in optical images of single Ag nanoparticles embedded in those tissue sections outlined in (**a**). Dashed circles outline the representative embedded individual Ag nanoparticles. Scale bar = 400 μm (**a**) and 4 μm (**b**). Reprinted with permission from Ref. [18]. Copyright (2007) American Chemical Society

toxicity, as shown in Fig. 8.10. Individual Ag nanoparticles exhibited size-dependent scattering colours, which enabled the differentiation of nanoparticle size during membrane transport and diffusion at the single-particle level. The membrane transport of Ag nanoparticles occurred through chorion pore canals (CPCs) and exhibited Brownian diffusion. The Ag nanoparticles were observed to passively diffuse into developing embryos via CPCs and to exert specific effects on embryonic development. Some kinds of abnormalities observed in living zebrafish were highly dependent on the concentration of Ag nanoparticles. The diffusion rates and accumulation of the nanoparticles in the embryos could be related to dose-dependent abnormalities. This study marked the first direct observation of CPCs in living embryos by microscopy. In addition, the Ag nanoparticles were used in living microbial cells to monitor changes in membrane permeability and pore size at the nanoscale in real time. The results indicated that Ag nanoparticles transported into and out of membranes and accumulated inside cells. Notably, the accumulation increased as the chloramphenicol concentration increased, which indicated that chloramphenicol enhanced membrane permeability. Compared with Ag nanoparticles, gold nanoparticles exhibited better biocompatibility and decreased toxicity in the same study. Thus, nanoparticles with specific scattering colours can provide a new way to investigate and reveal the molecular transport mechanisms in vivo to help elucidate related pathways.

In addition, gold nanoparticles (less than 50 nm) have been used in vaccine delivery because of their excellent biocompatibility and facile tracing. Sangyong Jon and his co-workers first reported GNP-based cancer vaccines that enabled the delivery of antigens to target cells [20]. This vaccine delivery method, which exhibited significant antitumour efficacy, could be tracked by noninvasive clinical imaging, as displayed in Fig. 8.11. This method provides a new direction for the application of plasmonic nanoparticles in cancer prevention and therapy.

8.4 Plasmonics Nanoparticles in Photothermal Therapy

As previously described, the scattering spectra of metal nanoparticles can be modulated by tuning their shape, composition, and size. These properties provide a means to kill cancer cells via photothermal cancer therapy through the use of plasmonic particles with scattering spectra in the near-infrared (NIR) region, where optical transmission through tissues is optimal [21, 22]. The previously described anti-EGFR-modified gold rods with aspect ratios of 3.9 were used for cell imaging and photothermal cancer therapy. After treatment with the nanorods, three types of cells, a nonmalignant epithelial cell line (HaCaT) and two malignant oral epithelial cell lines (HOC 313 clone 8 and HSC-3) exhibited different affinities to the GNRs. The malignant cell lines could be recognised clearly under the 80 mW red laser with a wavelength of 800 nm, as displayed in Fig. 8.12. These three types of cells showed the same morphology under irradiation with a 40 mW laser, whereas the nonmalignant cells began apoptosis when the laser power was increased to greater than 120 mW. The good selectivity and targeting of cancer cells observed in this study demonstrates that nanoplasmonic materials with surface plasmon bands in the NIR region can serve as bifunctional contrast agents for photothermal therapy.

Gold nanorods with NIR plasmon resonance bands were also utilised for the in vivo photothermal therapy of deep-tissue malignancies (subcutaneous squamous

Fig. 8.11 **a, b** CT images obtained before (Pre) and 48 h after injection of RFP/AuNPs (**a**) or CpG/RFP/AuNPs (**b**) into the right rear footpad (*yellow arrow*). We observed the right popliteal LN (*golden arrow*) after injection (**c**) optical images of the dissected popliteal lymph nodes taken 48 h after injection of RFP/AuNP and CpG/RFP/AuNP vaccines (**d**) silver-stained sections of the right popliteal LN confirm the presence of RFP/AuNPs and CpG/RFP/AuNPs. Reprinted with permission from Ref. [20]. Copyright 2012 WILEY-VCH Verlag GmbH & Co. KGaA, Weinheim

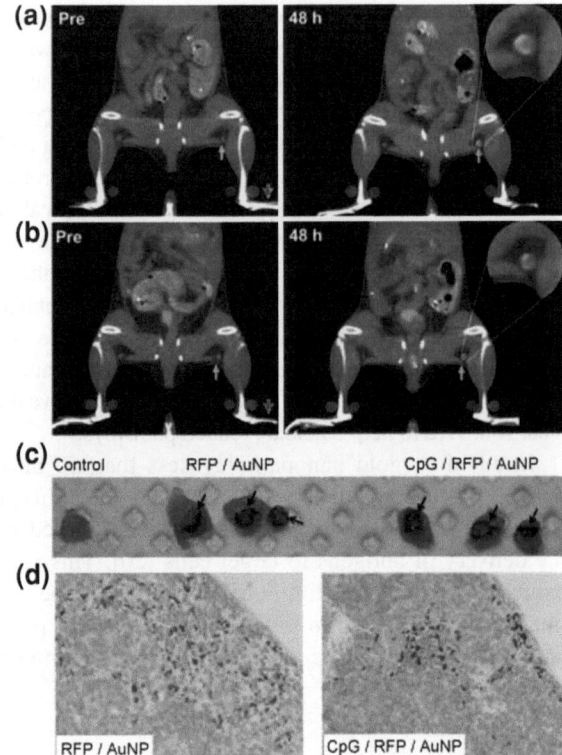

Fig. 8.12 Selective photothermal therapy of cancer cells with anti-EGFR/Au nanorods incubated. The circles show the laser spots on the samples. At 80 mW (10 W/cm^2), the HSC and HOC malignant cells are obviously injured while the HaCat normal cells are not affected. The HaCat normal cells start to be injured at 120 mW (15 W/cm^2) and are obviously injured at 160 mW (20 W/cm^2). Reprinted with permission from Ref. [22]. Copyright (2006) American Chemical Society

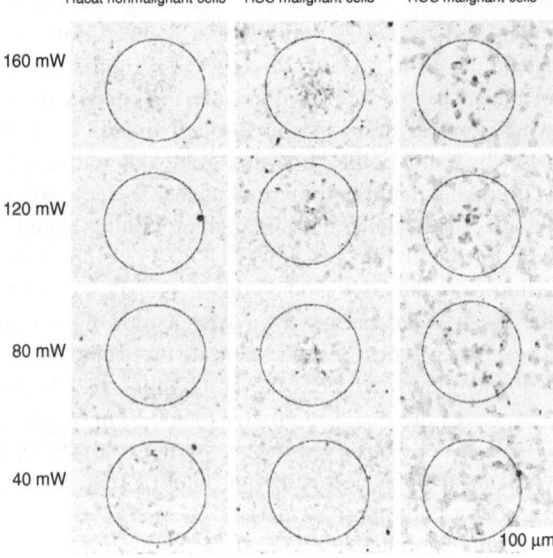

Fig. 8.13 NIR transmission images of mice prior to PPTT treatments. Inset shows intensity line scans of NIR extinction at tumour sites for control (*filled square*), intravenous (*filled triangle*), and direct (*filled circle*) administration of pegylated gold nanorods. Control mice were interstitially injected with 15 μL 10 mM PBS alone, while directly administered mice received interstitial injections of 15 μL pegylated gold nanorods ($OD_\lambda = 800 = 40$, 2 min accumulation), and intravenously administered mice received 100 lL pegylated gold nanorod ($OD_\lambda = 800 = 120$, 24 h accumulation) injections. Reprinted with permission from Ref. [23]. Copyright (2008) Elsevier

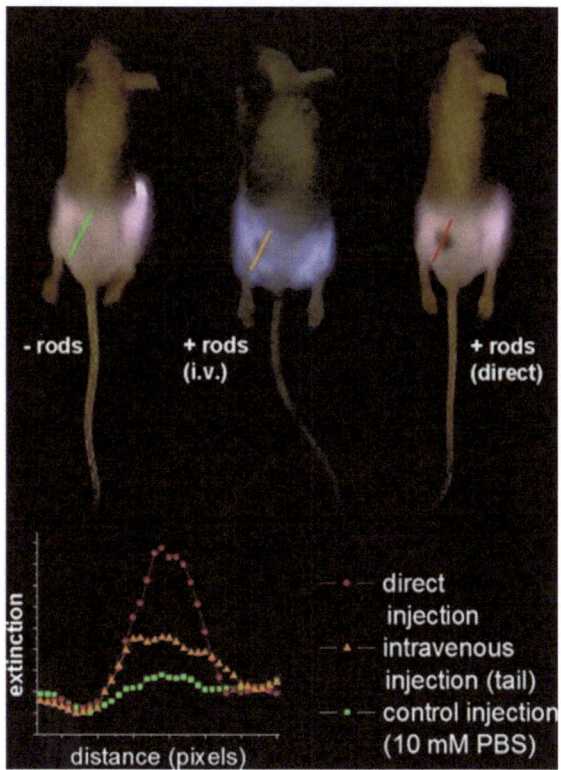

cell carcinoma xenografts, HSC-3) in nude mice irradiated by a small, portable, inexpensive NIR laser [23]. Significant decreases in cancer cell populations were observed after both the direct ($P < 0.0001$) and intravenous ($P < 0.0008$) administration of gold nanorods, as displayed in Fig. 8.13. The average inhibition of tumour growth for both delivery methods was observed over a 13-day period, and the resorptions of more than 57 % of the directly injected tumours and 25 % of the intravenously treated tumours confirmed the valuable potential of NIR plasmonic nanoparticles in thermal therapy applications.

To overcome the limitations of nanorods coated with the toxic surfactant cetyltrimethylammonium bromide (CTAB), a polyamidoamine dendrimer was used to replace CTAB on the nanorod surfaces. The polymer-modified gold nanorods were then conjugated to RGD to enhance their selectivity towards targeted cells for in vivo thermal therapy [24]. The results confirmed that these surface-functionalised gold nanorods could specifically target tumour cells and vascular cells inside tumour tissues in mice without cytotoxicity (Fig. 8.14). Under irradiation by an NIR laser, the RGD-GNRs exhibited clearly destructive effects on the cancer cells and solid tumours, and parts of the tumours in mice even disappeared. These high-performance nanoprobes exhibit great promise in applications for tumour targeting and thermal therapies for cells with overexpressed $\alpha_v\beta_3$ integrins.

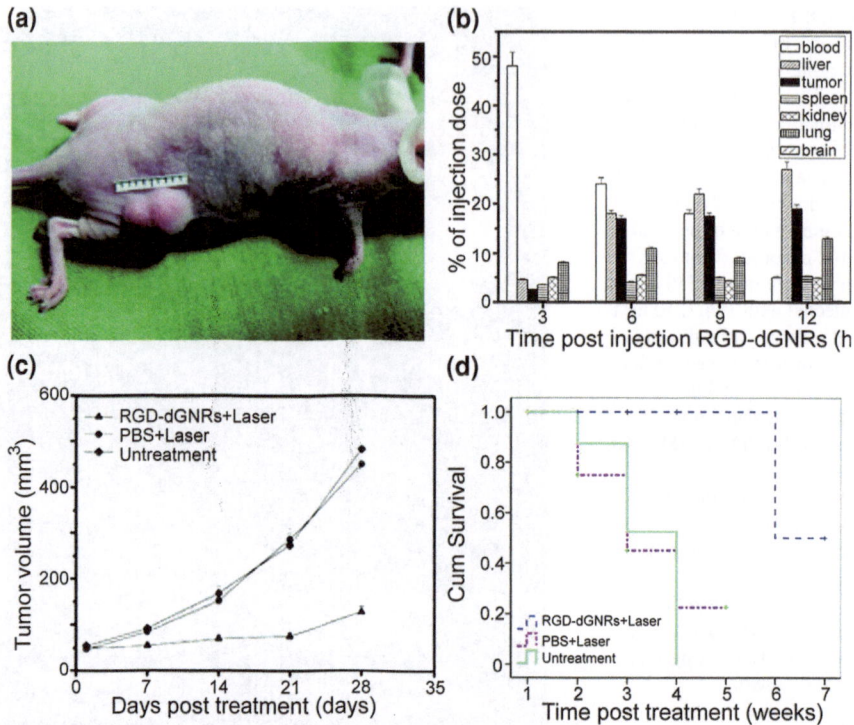

Fig. 8.14 Melanoma animal models, biodistribution of RGD-dGNRs, and survival data analysis of control and test group. **a** A375 melanoma mouse models. The tumour size can be calculated as $ab^2/2$ (a represents the longer dimension and b represents the shorter dimension of the tumour). **b** Biodistribution of RGD-dGNRs in mice after intravenous injection. Several time points after injection, gold amounts in tissue samples were evaluated by ICP mass spectrometry ($n = 3$). **c** Tumour size at different time points postirradiation of mice treated with RGD-dGNRs plus NIR laser (group 1); PBS plus NIR laser (group 2) or untreated control (group 3), $P < 0.05$ for group 2 or group 3 versus group 1. **d** Kaplan–Meier curve of the test group and control group, $P = 0.006$. Reprinted with permission from Ref. [24]. Copyright (2009) American Chemical Society

Similar to nanorods, metal nanoshells have variable optical plasmon resonance bands via tuning the shell composition and thickness [25, 26]. West fabricated gold–silica nanoshells that strongly absorb light in the NIR region to use in thermal ablative therapies for cancers in vitro and in vivo. The nanoshells were incubated with human breast carcinoma cells, and cell morbidity was induced when exposed to NIR light (820 nm, 35 W/cm^2). The resulted localized cell death was confined to the laser-nanoshell treatment area. As a control, cells without nanoshells exhibited no loss in viability after the same irradiation with NIR light. For the in vivo study, solid tumours incubated with metal nanoshells showed irreversible tissue damage at low powers of NIR light (820 nm, 4 W/cm^2, $\Delta T = 37.4 \pm 6.6$ °C) within 4–6 min; this damage was also confined to the tumour volume. Tissues heated above the damage threshold temperature showed distinct coagulation, cell shrinkage, and loss of nuclear staining, which indicated irreversible thermal damage. In contrast, the tumours without nanoshells exhibited lower average temperatures ($\Delta T < 10$ °C). These results suggest that metal nanoshells with good biocompatibility can be used for the thermal destruction of tumours.

In summary, due to the excellent properties of nanoplasmonics including strong intensity, stability, biocompatibility, nontoxicity, and facile modification, they can act as useful and sensitive nanoprobes in cell imaging and thermal therapy. The detection based on plasmon resonance paves the way to the exploration of metabolic process, immune reactions, and cancer healing which is meaningful in life science and iatrology.

References

1. Cognet L, Tardin C, Boyer D, Choquet D, Tamarat P, Lounis B (2003) Single metallic nanoparticle imaging for protein detection in cells. Proc Natl Acad Sci USA 100:11350–11355
2. Peeters S, Kitz M, Preisser S, Wetterwald A, Rothen-Rutishauser B, Thalmann GN et al (2012) Mechanisms of nanoparticle-mediated photomechanical cell damage. Biomed Opt Express 3:435–446
3. Xu L, Liu Y, Chen Z, Li W, Liu Y, Wang L et al (2012) Surface-engineered gold nanorods: promising DNA vaccine adjuvant for HIV-1 treatment. Nano Lett 12:2003–2012
4. Song Y, Xu X, MacRenaris KW, Zhang XQ, Mirkin CA, Meade TJ (2009) Multimodal gadolinium-enriched DNA-gold nanoparticle conjugates for cellular imaging. Angew Chem Int Ed 48:9143–9147
5. Connor E, Mwamuka J, Gole A, Murphy CJ, Wyatt MD (2005) Gold nanoparticles are taken up by human cells but do not cause acute cytotoxicity. Small 1:325–327
6. Rosman C, Pierrat S, Henkel A, Tarantola M, Schneider D, Sunnick E et al (2012) A new approach to assess gold nanoparticle uptake by mammalian cells: combining optical darkfield and transmission electron microscopy. Small 8:3683–3690
7. Patra HK, Banerjee S, Chaudhuri U, Lahiri P, Dasgupta AK (2007) Cell selective response to gold nanoparticles. Nanomedicine 3:111–119
8. El-Sayed IH, Huang X, El-Sayed MA (2005) Surface plasmon resonance scattering and absorption of anti-EGFR antibody conjugated gold nanoparticles in cancer diagnostics: applications in oral cancer. Nano Lett 5:829–834
9. Aaron J, Travis K, Harrison N, Sokolov K (2009) Dynamic imaging of molecular assemblies in live cells based on nanoparticle plasmon resonance coupling. Nano Lett 9:3612–3618
10. Austin LA, Kang B, Yen C-W, El-Sayed MA (2011) Plasmonic imaging of human oral cancer cell communities during programmed cell death by nuclear-targeting silver nanoparticles. J Am Chem Soc 133:17594–17597
11. Kang B, Mackey MA, El-Sayed MA (2010) Nuclear targeting of gold nanoparticles in cancer cells induces DNA damage, causing cytokinesis arrest and apoptosis. J Am Chem Soc 132:1517–1519
12. Huang X, Peng X, Wang Y, Wang Y, Shin DM, El-Sayed MA et al (2010) A reexamination of active and passive tumor targeting by using rod-shaped gold nanocrystals and covalently conjugated peptide ligands. ACS Nano 4:5887–5896
13. Hu R, Yong K-T, Roy I, Ding H, He S, Prasad PN (2009) Metallic nanostructures as localized plasmon resonance enhanced scattering probes for multiplex dark-field targeted imaging of cancer cells. J Phys Chem C 113:2676–2684

14. Sheridan C (2012) Proof of concept for next-generation nanoparticle drugs in humans. Nat Biotechnol 30:471–473
15. Kumar S, Harrison N, Richards-Kortum R, Sokolov K (2007) Plasmonic nanosensors for imaging intracellular biomarkers in live cells. Nano Lett 7:1338–1343
16. Yu C, Nakshatri H, Irudayaraj J (2007) Identity profiling of cell surface markers by multiplex gold nanorod probes. Nano Lett 7:2300–2306
17. Zhang L, Zhen SJ, Sang Y, Li JY, Wang Y, Zhan L et al (2010) Controllable preparation of metal nanoparticle/carbon nanotube hybrids as efficient dark field light scattering agents for cell imaging. Chem Commun 46:4303–4305
18. Lee K, Nallathamby PD, Browning LM, Osgood CJ, Xu XH (2007) In vivo imaging of transport and biocompatibility of single silver nanoparticles in early development of zebrafish embryos. ACS Nano 1:133–143
19. Xu X-HN, Brownlow WJ, Kyriacou SV, Wan Q, Viola JJ (2004) Real-time probing of membrane transport in living microbial cells using single nanoparticle optics and living cell imaging. Biochemistry 43:10400–10413
20. Lee IH, Kwon HK, An S, Kim D, Kim S, Yu MK et al (2012) Imageable antigen-presenting gold nanoparticle vaccines for effective cancer immunotherapy in vivo. Angew Chem Int Ed 51:8800–8805
21. El-Sayed IH, Huang X, El-Sayed MA (2006) Selective laser photo-thermal therapy of epithelial carcinoma using anti-EGFR antibody conjugated gold nanoparticles. Cancer Lett 239:129–135
22. Huang X, El-Sayed IH, Qian W, El-Sayed MA (2006) Cancer cell imaging and photothermal therapy in the near-infrared region by using gold nanorods. J Am Chem Soc 128:2115–2120
23. Dickerson EB, Dreaden EC, Huang X, El-Sayed IH, Chu H, Pushpanketh S et al (2008) Gold nanorod assisted near-infrared plasmonic photothermal therapy (PPTT) of squamous cell carcinoma in mice. Cancer Lett 269:57–66
24. Li Z, Huang P, Zhang X, Lin J, Yang S, Liu B et al (2009) RGD-conjugated dendrimer-modified gold nanorods for in vivo tumor targeting and photothermal therapy. Mol Pharm 7:94–104
25. Hirsch LR, Stafford R, Bankson JA, Sershen SR, Rivera B, Price RE et al (2003) Nanoshell-mediated near-infrared thermal therapy of tumors under magnetic resonance guidance. Proc Natl Acad Sci 100:13549–13554
26. Loo C, Lin A, Hirsch L, Lee M-H, Barton J, Halas N et al (2004) Nanoshell-enabled photonics-based imaging and therapy of cancer. Technol Cancer Res Treat 3:33–40

Chapter 9
Conclusions and Future Prospects

Localized surface plasmon resonance endows noble-metal nanoparticles with unique optical and physical properties. The strong absorption and scattering properties of plasmonic nanoparticles enable their use as sensitive and label-free sensors and probes. The LSPR band is dependent on the particle size, shape, composition, electron density of the nanoparticles, and their surrounding medium. Thus, researchers can modulate the plasmon resonance band by changing the morphology of the plasmonic particles such as nanorods, nanocore–shells, and nanostars, which exhibit excellent activity in catalysis, cell imaging, and photothermal therapeutic applications. In addition, by modifying nanoparticle surfaces, various sensors have been developed to detect changes in the refractive index of the surrounding medium. Through the application of dark-field microscopy, distinct breakthroughs in the scattering spectroscopy of single nanoparticles have promoted their development at the nanoscale. The elimination of averaging effects and ensemble properties enabled by single-particle measurements can enhance detection sensitivities by several orders of magnitude, even to the single-molecule level, thereby revealing the actual events that occur on individual particle surfaces [1, 2].

Currently, plasmonics have attracted considerable attention in the fields of catalysis, biology, optics, and chemistry. The next generation of plasmon resonance is a meaningful consideration. For example, noble-metal nanoparticles are readily modified by $-SH$ or $-NH_2$ groups, which enables the fabrication of special and intricate nanocomplexes with multiple functions through self-assembly. Such materials could be used in cell imaging, photothermal therapy, drug delivery or used for the highly sensitive, and selective characterisation of metabolic processes or biological interactions. In addition, the morphologies of nanoparticles can be tailored to tune the plasmon resonance band to a specific region for increasing detection efficiency and versatility. Notably, new fabrication methods are necessary in the investigation of plasmonics. The preparation of size- , shape- , and distance-controllable nanoparticle arrays will further the exploration and fundamental characterisation of particles and allow plasmon resonance oscillation to be finely tuned, thereby increasing reproducibility. In addition to the synthesis of

Y.-T. Long and C. Jing, *Localized Surface Plasmon Resonance Based Nanobiosensors*,
SpringerBriefs in Molecular Science, DOI: 10.1007/978-3-642-54795-9_9,
© The Author(s) 2014

functionalised plasmonic noble-metal nanoparticles, new materials with LSPR properties should also be developed. Recently, carbon nanotubes and silica were shown to produce LSPR scattering spectra [3–6]. The discovery of new plasmonic materials will extend both the basic theory and the applications of LSPR sensors.

LSPR arises from the resonance between surface electrons and incident light, which induces extremely fast electron oscillations and active energy areas. This photon–electron interaction provides a potential photovoltaic conversion mechanism to enhance energy transfer efficiency. Plasmonic materials coupled with TiO_2 have been used in the photocatalysis of carbon oxide, ethylene, and other organic pollutants through directional electron transfer between metal particles and TiO_2, which changes the electric potentials of both the semiconductor and the metal [7]. Meanwhile, LSPR-based spectroelectrochemistry and photocurrent detection are promising techniques for studying the basis of plasmon-enhanced electrochemistry. The high energy of metal nanoparticles facilitates molecular reactions on the particle surface, especially under excitation by incident light. These reactions can increase the electrochemical reaction activity and the Faraday electron transfer current. Furthermore, electrochemistry-enhanced plasmonic studies have confirmed that when a negative potential is applied to nanoparticles, the scattering light undergoes a blueshift and increases in intensity because of the increased electron density. Theoretically, the plasmonic enhancement of electrochemistry should be orders of magnitude because the thickness of the double layer is similar to the active electron area of the particles. However, the reports in the literature have thus far indicated enhancements of only a few-fold. We hope researchers will find an appropriate amplification system to improve the energy transfer ability.

Integrating LSPR with other advanced techniques may be the trend in the future. Dark-field microscopy coupled with surface-enhanced Raman spectroscopy offers a single-particle SERS detection approach for cell imaging [8]. Electrochemistry-modulated scattering spectra reveal the electron transfer mechanism of individual nanoparticles and provide more detailed information about reaction processes (Chap. 7). Furthermore, because nanopore electric methods are considered to be a third-generation DNA sequencing technique, the fabrication of functional plasmonic nanopores has become a popular topic in single-molecule detection [9]. Monitoring the simultaneous optical and electrical signals of single molecules translocating through a nanoscale pore overcomes the limitations of signal recognition in traditional nanopore approaches. In addition, plasmonics have exciting applications in signal transfer. Waveguides based on plasmonic structures have significant potential in optical and electronic communication applications [10–12].

In addition to sensor development, theoretical research is also important in guiding the field of plasmonics. Since Mie proposed the plasmon resonance theory on the basis of Maxwell's equations in 1908, many researchers have attempted to improve and extend the theory to arbitrary particles under different conditions. Predicting the plasmon resonance band of particles with various sizes, shapes, and compositions is critical to understanding the basis of LSPR. In addition, the ability to estimate the size or shape of single-plasmonic nanoparticles according to their LSPR spectrum is also very useful. We demonstrated a method for calculating the particle size of individual nanoparticles using dark-field microscopy, which provides an approach to achieve sub-100 nm resolution through conventional optical instrumentation [13]. Furthermore, the simulation of the active surface electron density effect on LSPR also remains a challenge.

In conclusion, nanoplasmonics provides a broad stage for the design of superior sensors in applications across numerous fields of study. LSPR-based observations, especially at the single-particle level, will undergo rapid development and improvement over the next several decades.

References

1. Li Y, Jing C, Zhang L, Long Y-T (2012) Resonance scattering particles as biological nanosensors in vitro and in vivo. Chem Soc Rev 41:632–642
2. Zhang L, Li Y, Li DW, Jing C, Chen X, Lv M et al (2011) Single gold nanoparticles as real-time optical probes for the detection of NADH-dependent intracellular metabolic enzymatic pathways. Angew Chem 123:6921–6924
3. Chou L-W, Shin N, Sivaram SV, Filler MA (2012) Tunable mid-infrared localized surface plasmon resonances in silicon nanowires. J Am Chem Soc 134:16155–16158
4. Wang F, Li C, Chen H, Jiang R, Sun L-D, Li Q et al (2013) Plasmonic harvesting of light energy for suzuki coupling reactions. J Am Chem Soc 135:5588–5601
5. Andolina CM, Dewar AC, Smith AM, Marbella LE, Hartmann MJ, Millstone JE (2013) Photoluminescent gold–copper nanoparticle alloys with composition-tunable near-infrared emission. J Am Chem Soc 135:5266–5269
6. Joh DY, Kinder J, Herman LH, Ju S-Y, Segal MA, Johnson JN et al (2010) Single-walled carbon nanotubes as excitonic optical wires. Nat Nanotechnol 6:51–56
7. Furube A, Du L, Hara K, Katoh R, Tachiya M (2007) Ultrafast plasmon-induced electron transfer from gold nanodots into TiO₂ nanoparticles. J Am Chem Soc 129:14852–14853
8. Austin LA, Kang B, El-Sayed MA (2013) A new nanotechnology technique for determining drug efficacy using targeted plasmonically enhanced single cell imaging spectroscopy. J Am Chem Soc 135:4688–4691
9. Ying YL, Zhang J, Gao R, Long YT (2013) Nanopore-based sequencing and detection of nucleic acids. Angew Chem Int Ed 52:13154–13161
10. Grzelczak M, Mezzasalma SA, Ni W, Herasimenka Y, Feruglio L, Montini T et al (2011) Antibonding plasmon modes in colloidal gold nanorod clusters. Langmuir 28:8826–8833
11. Février M, Gogol P, Aassime A, Mégy R, Delacour C, Chelnokov A et al (2012) Giant coupling effect between metal nanoparticle chain and optical waveguide. Nano Lett 12:1032–1037
12. Solis D Jr, Willingham B, Nauert SL, Slaughter LS, Olson J, Swanglap P et al (2012) Electromagnetic energy transport in nanoparticle chains via dark plasmon modes. Nano Lett 12:1349–1353
13. Jing C, Gu Z, Ying Y-L, Li D-W, Zhang L, Long Y-T (2012) Chrominance to dimension: a real-time method for measuring the size of single gold nanoparticles. Anal Chem 84:4284–4291